W9-BLF-966

Demanding Clean Food and Water

The Fight for a Basic Human Right

Demanding Clean Food and Water

The Fight for a Basic Human Right

Joan Goldstein

Plenum Press • New York and London

ISBN 0-306-43570-5

© 1990 Joan Goldstein
Plenum Press is a Division of
Plenum Publishing Corporation
233 Spring Street, New York, N.Y. 10013

Printed in the United States of America

To the memory of my friend,
Thomas M. Johnson

Preface

When the leaders of Princeton University's Earth Day 1990 celebration had invited me to speak at their Environmental Teach-In, I accepted with enthusiasm. I did wonder what message I would offer. It would be one I hoped would spark the interest of young students, most of them under the age of twenty and not even born at the time of that historic 1970 event—the world's first "Earth Day." I remembered it quite well and shared my memories of two decades past—of meeting young students of another era, men and women garbed in the costume of their predecessors, the "flower" and "peace movement" children. The Earth Day students gave each passerby a single blossom—a daisy, as I recall—to invoke a symbol of our good though troubled earth. My 1990 Princeton audience listened earnestly—I could have been describing the Civil War, for all these moments in time predated their very existence. The topic for this particular Teach-In was titled "How Far Have We Come since Earth Day 1970," and a most distinguished panel of experts highlighted by Princeton professor Richard Falk, Richard Ayres of the Natural Resources Defense Council, biology professor Steve Hubbell, an Arco Chemical representative Steve Brown, an Assistant Commissioner of the New Jersey Department of Environmental

Protection, and myself included, made an attempt to bridge the gap in time. Yes, there had been advances made in those two decades, especially in the realm of legislation. But another concern came to light that same evening. The audience of students wanted to know what they could do now, at this present time, to save their endangered environment. We attempted to answer that question as well. Many of our suggestions involved both small, individual, behavioral changes, and also large scale institutional shifts in focus. It was my thrust to focus on changing our agricultural practices.

I encountered that same nagging question a day or two later when I spoke before an Earth Day audience of three-hundred West Windsor high school students in New Jersey. "What can we do?" queried a very bright and articulate president of the senior class. It was rewarding to see the interest and concern.

We are, it seems, at a crossroad in this business of preserving the environment. There is no question that we have stimulated a national if not international interest in the many threatening environmental problems. The mining and coordination of solutions is ahead of us still. What is necessary now is to foster the leadership, the information, and coordination to move Earth Day from the status of a celebration to that of a continuous, working objective for our society. If we recall in the Bible, not even the earth was created in one day's time. We can ask no more of Earth Day than to awaken the public to the most formidable challenge of our time.

Joan Goldstein

Princeton, New Jersey

Acknowledgments

This book was made possible by the help and dedication of many others. In particular I wish to thank Carolyn T. Cohen, International Training Specialist, Program Development Center, Animal Plant and Health Inspection Service of the U.S. Department of Agriculture, and Gary Cunningham, Program Manager, Biological Control Operations, Plant Protection and Quarantine, APHIS, U.S. Department of Agriculture; Dr. Richard G. Ames, Epidemiologist with the California Department of Health Services; and Dr. Michael A. O'Malley, Medical Coordinator with the California Department of Food and Agriculture Pest Management, Environmental Protection and Worker Health Safety branch. Enormous help and encouragement came from Maura Cantrill, Barbara Cantrill, Elisabeth and David Hagen, Drs. Bogdan Denitch and Craig Humphries, my former professor Dr. Sylvia Fava, my brother Dr. Michael J. Goldstein, political/economic officer for the Canadian Consulate General David Biette, Warren Doering, Lucile Bond, and Melisa Gusciora. I particularly wish to acknowledge the professional working relationship and invaluable assistance offered by my editor at Plenum, Linda Greenspan Regan. It was her keen eye for the philosophical and substantive content of this book which led to its auspicious completion.

Contents

CHAPTER ONE

The Start of the Chemical Age

Paradise Found and Lost

Coming upon an apple tree on the edge of the wilderness is an extraordinary sight. The trunk and branches are horribly gnarled and twisted, as if scarred by the birth of the fruit. The small, rounded, pungent fruits are green in cast and sour to the taste. Beneath the knotted branches of the tree lie the fallen apples. A heavy, sweet perfume exudes from the fermenting and bruised, irregular shapes. These little, wild apples are a far cry from the ones we buy in the supermarket today. Arranged in orderly displays, large, brilliant red, perfectly round or oblong, grown without the boring of worms and chemically ripened to perfection, the apples we consume are made perfect by the application of pesticides.

Originally, pesticides were designed to control or kill undesirable pests. This soon led to the development of insecticides which control, or literally kill, insects (the

Latin suffix "cide" refers to the act of killing). As their value became apparent for agricultural uses, other chemicals were developed under an umbrella category known as "pesticides." These included herbicides to control weeds; fungicides to control fungi such as mold and mildew; miticides to do away with mites; and rodenticides to control rodents. Apples alone are sprayed with the fungicides Captan, Mancozeb, and Folpet, plus the growth regulator daminozide (trade name, Alar), although this last one has been a subject of recent controversy. In 1989, there were many heated discussions about the spraying of Alar, a chemical which allows the quick ripening of rosy, red fruit. More and more people are becoming aware that our fruits and vegetables have arrived at this state of perfection through the immoderate, even excessive, application of chemicals. This process might be proving a serious risk to our health as well as to our rapidly dwindling supply of clean water. Through daily consumption of food and water, people are exposed to pesticides. Even if the exposure is at low levels, it is chronic exposure nevertheless. Food is the major source of pesticide absorption, but pesticides are also absorbed from drinking water, from contaminated air, and through contact exposure with skin. Virtually everyone in the United States, including our children, has some pesticide residue in his or her tissues.[1] How did this happen?

Only a hundred years ago, the magnificence of our rich, abundant, and largely unpolluted American countryside inspired a nineteenth-century Old World composer, Anton Dvorak, to write his "New World Symphony" in tribute to our unspoiled land. From a railroad car, crisscrossing the vast American landscape, the Bohemian composer delighted in the beauty of the farmlands, rivers, and streams that made up what was still referred to as "The New World."

In one hundred years, we have travelled in time and in space: The New World is aging, somewhat badly, and the luster of a pure environment has been marred. A recent report published by a public interest group known as The Natural Resources Defense Council offers some alarming news about our abundant food supply. This report, called "Intolerable Risk," tells us that

> our nation's children are being harmed by the very fruits and vegetables we tell them will make them grow up healthy and strong. These staples . . . routinely, and lawfully, contain dangerous amounts of pesticides, which pose an increased risk of cancer, neurobehavioral damage, and other health problems. . . . Little is being done by the government to protect children from the intolerable risk to their health posed by pesticide residue in food.[2]

Though the federal agency primarily concerned with environmental protection, the EPA (Environmental Protection Agency), agrees to some extent with the conclusions offered by the NRDC (Natural Resources Defense Council), it questions the degree of risk associated with ingesting pesticides over a long period of time.

What can consumers conclude from the often contradictory statements made by experts? Should we as consumers turn our backs on our enormously abundant food supply—the red apples, lush strawberries, oranges, potatoes—all the fruits of the earth it would seem, simply to avoid the dreaded pesticide residue that could conceivably, over time poison our bodies? Can we deal with the dilemma of abundant and attractive food, and the ill effects produced by the very pesticides that apparently made them possible. Do we have any choice in the matter?

To search for the answers to these questions we need
to look into a historical rearview mirror. What was done in
the past is certainly the key to the present. If we went back
in time, and observed the food producers of past eras, we
would find a very different agricultural landscape—so
different, in fact, that we would barely recognize it as our
own. For it was only about 50 years ago that the whole
system of supplying food and water to the entire nation
was accomplished without the aid of massive doses of
synthetic chemicals. That does not mean that agriculture
was ever an easy or sure endeavor before the introduction
of modern chemicals. The capriciousness of nature was
always a force to contend with: storms, droughts, drench-
ing rains and floods, and the dreaded invasion of pests
punctuated all eras. Wheat, potatoes, corn, tomatoes, apples, peaches, virtually everything that grew in or up from
the ground was subject to disfigurement or total devastation by vigorous armies of insects. In addition came the
problem of choking weeds and ravenous rodents that had
to be discouraged somehow from completely destroying
the crops. It's no small wonder that when chemicals were
introduced for agricultural use following World War II, in
the mid-1940s (though some had certainly made their way
into crop dusting in the 1930s), they were seen as "miracle
workers" that would revolutionize farming. And, in fact,
they did just that. Once pesticides entered the scene at the
onset of the mechanized age, the face of food production
changed forever. It is not surprising that for many years,
farmers, the food industry, the government, and the public all viewed these changes in a highly positive light. The
use of pesticides was considered a significant step forward in improving our quality of life. It is only now, some
50 years down the road, that we are beginning to gather

some perspective on what we did and did not accomplish in those postwar years.

THE CHEMICAL AGE BEGINS: PESTICIDES AS PANACEA

To understand the dogged faith that the agricultural business community has placed on the "miracle of pesticides" for the last 50 years, we would first have to see the problem of pest control as they saw it—in its most flagrant form. Nothing could be more dramatic in action or in severity of damage than a serious invasion of grasshoppers, or locusts (they are actually the same species).

A questionable act of nature, plagues of locusts have been on record since biblical times. In the story of Exodus, we remember Moses negotiating the Israelites flight from slavery in Egypt. Demanding release of his people, Moses threatens the Egyptian Pharoah with the almighty's intervention—the 10 Deadly Plagues, of which the unleashing of locusts was number 8. Then, as now, a forceful army of winged creatures could denude food crops of all kinds, fields of grain particularly, and leave starvation and famine in their wake. It is obvious why locusts achieved the distinction of being labelled a "plague" even in those ancient times.

The first locust outbreak of serious proportions in the twentieth century developed in the U.S. farmbelt of South Dakota, Nebraska, and Iowa in 1931. Since recorded time, locust outbreaks seem to occur in a cluster for five or six consecutive years and then disappear again for another fifty. According to John J. Schlebecker, an agricultural historian, "localized hordes (of locusts) came in 1932 and

1933, increasing in 1934. Destruction of crops could be 100 percent on a vast acreage. Damage increased in 1934, decreased in 1935, and then hit a peak in 1936."[3] By destroying every leaf, the locusts laid the land open to wind and erosion, contributing, along with some unwise farming practices, to the devastating dust bowl drought of the 1930s depression years.

Earlier, in the nineteenth century, a similar outbreak of locusts is recorded from the American farmbelt. This devastating event arrived some 50 years before the damage during the Great Depression mentioned above. Using computers, the Department of Agriculture has only just begun to track the periodic patterns of locust infestations, so precise patterns in time are not actually known as yet, though some historical material exists. For example, an 1873 locust infestation that continued for six years was reported upon and vividly described in O. E. Rölvaag's classic 1927 novel, *Giants in the Earth*. Though Rölvaag had written the novel 50 years following these events, at the time he was a professor at St. Olaf College in Minnesota and was well able to collect historical records and oral histories from farm families. So the novel has historical relevance as well as dramatic impact. Thus, as we read in *Giants in the Earth*, hordes of locusts nearly destroyed a tiny 1870s Norwegian settlement in America. By examining *Giants in the Earth*, we get a very accurate description of what it is to experience a locust infestation. Rölvaag's sturdy and stoic homesteading immigrants had been surviving the hardships of frontier existence in the vast and desolate Western great prairie. After years of struggle, they are finally ready to harvest their first successful crop of wheat. The locusts suddenly appear. Per Hansa, the author's tragic hero, standing in the fields with his neighbors, looks up to see a strange and chilling phenomenon on the horizon moving in waves and rapidly approaching their direction:

"Off in the western sky he caught sight of something he couldn't understand—something that sent a nameless chill through his blood. . . . Could that be a storm coming on?"

The immigrants stand spellbound, watching in terror and helpless to stop the "ominous waves of cloud." It advances with terrific speed, "breaking now and then like a huge surf and with the deep, dull roaring sound, as of a heavy undertow rolling into caverns in a mountain side." Descending upon the two lonely farm families, the winged army, millions of locusts, set about to ravenously destroy everything in their path:

> And now from out the sky gushed down with cruel force a living pulsating stream, striking the backs of helpless folk like pebbles thrown by an unseen hand . . . it chirped and buzzed through the air; it snapped and hopped along the ground; the whole place was a weltering turmoil of raging little demons . . .
>
> "Father!" shrieked Store-Hans through the storm. "They're little birds—they have regular wings!" [The child cannot tell that these winged creatures are large insects, and the father replies slowly and solumnly,] "This must be one of the plagues mentioned in the Bible." [The narrator later continues:]
>
> After that it raged with unabated fury throughout the years 1874, 75, 76, 77 and part of 78; then it disappeared as suddenly and mysteriously as it had come. The devastation it wrought was terrible; it made beggars of some, and drove others insane. . .[4]

Though the plague of locusts seems to appear only after long years of dormancy, the devastation it reaps is remembered for decades afterward. Schlebecker wrote in 1975 that the last great locust plague hit the United States in 1939, when grasshoppers stripped Colorado, Wyoming, Nebraska, Iowa, South Dakota, Montana and North Dakota clean. "Nothing like that would ever recur again," comments Schlebecker in the mid-1970s, "because of the discovery of inorganic insecticides, of which DDT led the way."[3] So it was pesticides that were to provide the "miracle" solution to end the scourge of locusts once and for all. Unfortunately, that did not happen. DDT (dichloro-diphenyl-trichloroethane), which will soon be discussed, was not used on grasshoppers in the United States to any great extent, though other insecticides were tried. And it is true that we failed to see any sign of locusts in recent times— that is until the summer of 1988, and the following very wet summer of 1989.

Spring rains, widespread storms, and the subsequent flooding brought about ideal breeding conditions for insects. In Minnesota, officials called the 1989 grasshopper infestation the worst in 50 years; and in some townships, up to 80 grasshoppers per square yard were found—ten times what would be considered crop-threatening numbers. Again, as in the 1939 infestation, parts of North Dakota, Nevada, South Dakota, Wyoming, and Utah witnessed swarms of grasshoppers cutting fields to the ground in just a few short days.

Concurrently in 1988, the Department of Agriculture issued a "Pest Alert" when a most unusual pest situation developed in the Caribbean as a result of Hurricane Joan. The desert locust, endemic to Africa and the mid-East, appeared to have been "blown in" by Hurricane Joan and had reached the shores of Puerto Rico and Saint Lucia in the British Virgin Islands. Cargo shops, airlines, and ports of entry were alerted to search aircraft and vessels for the hitchhiking, unwelcome pests. Fortunately, the climate was not altogether conducive to the desert locusts' survival in the Caribbean, though they wreaked havoc to the food supply in Africa that same year. Unfortunately, once an infestation has begun, pesticides can only be applied to contain the damage, and this is done by spraying against the swarms of new generations of insects; but at this point we can certainly not control the onset of the locusts. Contrary to Schlebecker's prediction that we would never again encounter grasshopper infestations due to the use of pesticides, scientists in the Department of Agriculture now acknowledge that they will probably never be able to stop the infestations from occurring.

From this recent, fairly dramatic example of insect infestation and its potentially disastrous effect upon food crops and food supply, we can begin to understand why those engaged in the business of agriculture not only endorse but rather embrace the use of modern pesticides to solve the age-old problem of insect infestation and crop destruction. But in reality, although the power of pesticides has produced some startling short-term successes against pests, they have been followed by dangerous long-term effects. These effects have included the generation of tenacious and resistant strains of genetically altered insects and plants whose destruction can only be controlled by escalating the dosage of newer, more lethal chemicals in an ever spiralling cycle. The complexity of this problem is staggering. While the new voracious insects reproduce succeeding generations of chemically immune descendents, the workers who handle pesticides, the public who consume products tainted with pesticide residue, and the precious water supply that receives chemical runoffs are all part of the risk for an apparently better quality of life. That better quality of life depends on the innundation of about one billion pounds of herbicides, insecticides, fungicides, and all other toxic chemicals per year, creating a totally chemically dependent food supply.

Hence, the chemical age and its offspring, the "miracle" pesticides, originally enjoyed a far better reputation than they had a right to claim. As with the grasshopper infestation, pesticides were used to control pests but did not ultimately contain the problem; rather, in the long run, these chemicals introduced greater dilemmas for our ecology. These serious ecological concerns were first introduced to the public mind in the early 1960s by the biologist and writer, Rachel Carson. Her monumental contributions were earth shattering at a time when the world was applauding the virtues of pesticides. Rachel Carson alerted and alarmed the public against the use of pesticides by bringing to light its dangers to the ecosystem, the cycle of nature.

But there were certainly other reasons why those in agriculture found it necessary to persist in pesticide use long after the damaging effects were exposed by Carson. The subsequent use of pesticides for 20 years following the publication of her book *Silent Spring*[5] was based on changes that were taking place in agricultural land use. These new methods significantly altered the scale and scope of farming practices in America, and eventually in every part of the world.

A CHANGE IN FARMING FROM AGRICULTURE TO BIG BUSINESS

A major change in farm practices which began in the 1930s contributed to the rising use and ultimate dependency upon "miracle pesticides" by the farming community in the United States and in Europe. These changes began with the massive consolidation of agriculture.

In the Great Depression years of the 1930s, small subsistence farming virtually disappeared from the landscape. A failed economy led to bank foreclosures, mostly on small-scale farm families who could least afford to survive in the downturn. Moreover, an ecological disaster added to the loss of arable farmland. The "dust bowl," a term used to describe the desertification of the American Southwest in those early depression years, forced the poor, independent farm family from their homes and led to the now legendary drift westward to the "land of milk and honey," California. In the rich California cotton fields and fruit groves, hundreds of thousands of homeless and disenfranchised farm families became the migrant farm workers and fruit pickers who would number among the earliest victims of pesticide exposure.

What drove sharecroppers from their Oklahoma, Arkansas, and Texas panhandle farms to California were three of the most powerful economic and social forces. The first was the mechanization of agriculture. Tractors, combines, automation, self-propelled equipment, and hydraulic lifting equipment all came to replace the labor intensive work of farmers. As long as energy was cheap, and once the initial investment had been made, mechanized equipment could prove less costly than sharecroppers or itinerant workers. So strong was this trend toward mechanization of farms that excluding garden tractors, the number of tractors on U.S. farms rose from 1,567,000 in 1940 to 2,354,000 in 1945, a total increase of 50%, or 10% a year.[6]

The second social and economic force to influence the scale of farming was the encroachment of large agricultural interests. The new owners of agricultural land included distant corporations and conglomerates who purchased abundant parcels of land as it became more and more apparent that food production could be redesigned into a thriving industry.

The third powerful economic and social force was the failure of the land itself to sustain life. The desertification of the American Southwest turned arable farmland into dust. Leached of valuable minerals, through unwise farm practices, the drought, and the resulting dust bowl, the scarred and overworked soil was unable to support life. All three of these forces inadvertently conspired to erase small-scale farming from the map of this country. As small-scale farms fell under the weight of drought conditions, they were bought from the banks by more affluent citizens. Eventually, agriculture emerged as something we now call agribusiness. With the introduction of mechanical helpers, it became feasible to combine larger tracts of acreage, a feat that would previously have required a substantial number of field workers. As a result of the development of mechanical devices, like the tractor, fewer workers were needed. Farms soon became more and more massive.

In order to consolidate the land, it was necessary to dismantle the age-old tenant farming system. In tenant farming, which can be traced back to the Middle Ages, a small farm family was given a tract to cultivate. In return, the family contributed half or more of its crops to the "owner man." This sytem worked as long as labor intensive, smaller-scale farm work was the only choice for cultivating the land. With the advent of high technology, machines were able to do the work of people, and the tenant-farming system was abruptly dismantled.

John Steinbeck's *The Grapes of Wrath* poignantly depicts these changes. Among those migrant tenant farmers were the Joad family, fictional characters created by author Steinbeck in his American gothic novel. Published in 1939, this novel startled American readers with its depiction of poor evicted sharecroppers. The "owner men" come late one evening, Steinbeck wrote, to dismiss and evict their tenant farm families who have lived on the land for generations:

> At last the owner men came to the point. The tenant system won't work any more. One man on a tractor can take the place of twelve or fourteen families. Pay him a wage and take all the crop. We have to do it. We don't like to do it. But the monster's sick. Something's happened to the monster.
>
> [This reference to the "monster" alludes to the bank's practice of profit making from foreclosures on farms.]
>
> But you'll kill the land with cotton.
>
> We know. We've got to take cotton quick before the land dies. Then we'll sell the land. Lots of families in the east would like to own a piece of land.[7]

This last passage refers to the second major change that occurred with the consolidation of agriculture—that is, the decision to cultivate exclusively one or perhaps two crops, cotton for example, instead of diversifying crops. The diversification and rotation of crops is a better form of agriculture because in the long run it is healthier for the vitality of soil. This decision toward single-crop cultivation was yet another important step which led to the heavy dependence on pesticides. Not that agronomist George Washington Carver had not warned us about the necessity of crop rotation, particularly with cotton. But it was thought more profitable from the point of view of cultivating, processing, and marketing to cultivate one major crop over a large acreage. Though it was indeed profitable in the short term, it opened up a hornet's nest of new problems. Concentrated crops brought their accompanying pests and they came in greater numbers than before. Tenant farmers, after all, had been subsistence farmers who diversified and varied the cycle of crops, cultivating food for themselves as much as meeting the business interests of the landowners. This benefited the farmers, enriched the soil, and kept the pest population under control all at the same time. Not only did tenant farming soon evaporate, but small-scale independent farm owners also lost out during the depression. If they couldn't pay their debts, the bank took over, auctioned off the farm and sold the land to the expanding landholder interests. Inadvertently, this large-scale cultivation of crops led to an escalation of the pest population problem.

Thus, with land consolidation and single-crop cultivation, there was a concurrent increase in the pest population specific to that one crop. For example, whereas grasshoppers are the exception and ravage everything in sight, most insects are selective feeders. They usually eat only one specific crop, or even one part of the plant. (A good example is the Russian wheat aphid, which only feeds on wheat and barley, and can be found in large numbers in any wheat field.) A pest population that came to feed upon one specific crop, its dietary mainstay, now had acres and acres of food available to it. The greater the food supply, the greater the potential to sustain insect life. And often, there were no other competitors for that food source. Predators that might have existed through diversified crop development were no longer on the scene. By the time that pest population and the acreage of working land had increased substantially, only the most dramatic actions could be applied to stem the tide. That meant labor-intensive measures. As every farmer calculated, labor-intensive work was costly. (So too were the newly developed pesticides; though later on, in greater bulk use, the price would eventually drop.) Farmers embraced pesticides as the most efficient solution. What they did not know about—could not know about—at that time were the secondary effects of this widespread spraying of pesticides—effects on landscape, crops, water, wildlife, human ecology, and yes, even on the pests that they had originally intended to destroy.

Following the changes brought about during the Great Depression, one more event was to occur before the modern development and widespread use of pesticides became an institutionalized practice. That event dramatically changed our way of life for the next 50 years. Our modern society was shaped by one catastrophic circumstance known as the Second World War.

WAR AND CHEMICALS: THE CREATION OF PESTICIDES

Both World Wars were responsible for spurring on the research and development of chemicals that would ultimately become the basis for pesticides used for peacetime agriculture.

During World War I (1914–1918), we witnessed the horror of modern high-tech warfare. Phosgene, used today to produce chemical herbicides and insecticides, was originally developed for use in chemical warfare. It was the most pernicious ingredient of poison gas in World War I.

When dropped on the enemy, mustard, or poison, gas was inhaled into the lungs, destroying the tissues of the respiratory system. This poison was used openly and without restraint by all sides during the conflict. In battlefields, the soldiers' only protection from inhaling clouds of sickening gas was to don the primitive gas mask—the round, goggle-eyed, elephant-trunked nose and mouth cover that was issued to each foot soldier. But these masks were hardly sufficient protection. Following the war, poison gas was outlawed for use by the League of Nations, the antecedent of the United Nations; however, recent reports from the Iran–Iraq conflict and Angola suggest it is still in use.

Early Compounds: The Birth of DDT

Focused on the development of chemicals for warfare, Germany developed organophosphorous compounds, the precursors of our current pesticides, as potential nerve gases. As early as 1874, DDT (short for dichloro-diphenyl-trichloro-ethane) was first synthesized by a German chemist. No one found a practical use for it until much later, in 1940, when a research group in the Swiss firm of J. R. Geigy discovered the effectiveness of the compounds in killing potato beetles and clothes moths. It also had a truly humane public health use that was discovered in the concluding years of the Second World War.

Toward the end of World War II, by mid-1944, DDT was standard issue for checking the spread of killer, insect-related diseases, such as typhus and malaria. These diseases were rampant in both the Allied forces and the liberated, formerly imprisoned civilian population. Lack of proper sanitation led to lice infestation, and that, in turn, brought about the deadly disease typhus. One can still catch glimpses in old black-and-white World War II newsfilms of the DDT-dusting of emaciated survivors of death camps.

At the time, it was impossible to think of DDT, or of any other modern chemical, as anything but the most benign of modern innovations. Even Rachel Carson, who was perhaps the first scientist to identify DDT as the most lethal of pesticides, acknowledged the relatively safe and humane uses it served. But DDT's relative safety was due largely to its method of application. DDT in powder form is not readily absorbed through the skin. When dissolved in oil (because DDT itself is fat soluble), as it usually is today, DDT is highly toxic.[8]

Toxic or not, DDT was welcomed with open arms by the expanding U.S. agricultural system for one very important reason. This chlorinated hydrocarbon controlled a wider range of external pests than any other insecticide. As a result, farmers applied it liberally across fields and streams in the United States and Great Britain. During the war-torn years of World War II, food production in Europe and Asia was considerably diminished. Across the Atlantic Ocean, the United States, unscathed by bombings and dislocation of farming communities, had risen to the stature of a major food producer. The U.S. armed forces had to be fed, but so did most of the embattled Allied nations that were no longer able to cultivate their own war-torn land. By 1942, the United States was exporting 23 percent of its cheese and dried skim milk, 7.2 percent of its butter, 10 percent of its dehydrated eggs, and smaller percentages of other food to its allies in Europe.

The demand for greater and faster food production

was accomplished with the newly consolidated farmland and high-tech mechanized equipment, despite the fact that there were at least 10 percent fewer farm workers (some of whom were in the armed forces). With larger farms and pressure for high production, the farmers demanded greater quantities of pesticides as well. The new system of large-scale, single-crop agriculture was already showing signs of greater pest problems, and the use of chemicals as an essential tool of agriculture was just beginning to take root in practice.

WORLD WAR II AND THE DISCOVERY OF HERBICIDES

As insecticides control insects, so do herbicides control weeds. Herbicides are also a product of chemical warfare research, and were stumbled upon by accident while researchers were studying plant growth hormones.

As early as 1939, scientists discovered that synthetic hormones could make ripened fruit hang onto trees longer and thereby hasten the ripening and coloring of fruit. But they also noticed that too much of these various hormones could damage and even kill plants. By 1941, with the idea that synthetic hormones could make good weed killers, 2,4-D (dichlorophenoxyacetic acid) was born.

By 1945, the very first year of the end of World War II, the herbicide 2,4-D became yet another chemical panacea for U.S. and European agriculture. In only one year, by 1946, production of 2,4-D increased nearly 500%; and by 1950, farmers applied the new herbicide to 30 million acres of cropland in the United States. Later, because it was effective at killing broad-leafed plants, 2,4-D was used by the military in Vietnam under a different name—Agent Orange. Here, it was argued, its value as a military weapon was in denuding the thick jungle growth where the enemy, in guerrilla fashion, hid from U.S. helicopters.

In summary, the fact that pesticides were spawned out of research in chemical warfare but applied and used as a peacetime agricultural panacea led to a mostly positive view of them in the public mind. Was the use of chemicals in food production harmful or was it helpful? During the transition years, from the 1930s to the 1950s, while farming practices changed from agriculture to agribusiness, chemicals played an ever-expanding role in massive food production. The portrayal of pesticides as helpful to our changing way of life was almost unquestioned. Concrete evidence existed and proved that our nation's crops were growing stronger and more abundant than ever before since the use of pesticides. The other side of the argument, the fact that the use of pesticides could be harmful, was less audible and certainly lacking in cumulative evidence. But still, in retrospect it might seem disquieting to realize that chemicals developed for warfare, for the very act of killing, were being applied so liberally to the foods we ate and, inadvertently, to the water we drank. The case for the potentially harmful effects of pesticides was yet to be made, since the evidence had only just begun to surface. We would journey forward into the 1960s before the public would even begin to suspect the dangers inherent in this new way of life.

CHAPTER TWO

A Voice in the Wilderness

Fairly early on, the potential harmful effects of pesticide use was brought to the attention of a very small, select group of scientists. As far back as 1938, at an entomology conference held in Berlin, Germany, A. J. Nicholson presented a paper on the indirect effects of pesticides.[1] As might be expected, the few farm workers who sprayed pesticides did so without knowledge of the potential harmful effects on their own health. Without necessary protection of their skin and respiratory systems, the farm workers could be poisoned due to their heavy and continuous exposure to pesticides. But the number of victims reported when an incident was first noted was largely overlooked. Pesticide use at the time was certainly not widespread enough to be noteworthy, nor its effects statistically significant. Later, following the war, in 1945, an article in the *Atlantic Monthly* written by V. B. Wigglesworth raised some serious questions about the use of DDT and its possible effects on the balance of nature.[2] As far as anyone was able to determine, the article had little or no effect upon the public or the developing chemical industry.

It was not until some 15 years later, in the very early

1960s, that concern for the effects of chemicals on the total ecosystem was laid open before the public. The public responded to the imminent crisis for the first time and did so with considerable alarm. The idea that these new, miracle chemicals had a dark side and could cause inadvertent yet dangerous changes to the very balance of nature had never been considered by the public. After all, these were miracle pesticides: powerful chemicals that had revolutionized the practice of modern farming and had made the production of an abundantly rich food supply possible. Suddenly, in the booming era of the 1960s, a biologist and writer named Rachel Carson became the first to challenge America's faith in chemicals.

SILENT SPRING: A VOICE IN THE WILDERNESS

Rachel Carson first shook the public's consciousness with a series of articles published in the *New Yorker* beginning on June 16, 1962. Later, these same startling ideas appeared in her now classic book, *Silent Spring*.[3] Dr. Rachel Carson articulated her examination of our environmental dilemma in a style that the public could easily comprehend. Her writing was clear, free of oblique scientific references, and further enhanced by delicate drawings of trees and flowers. Not only were her ideas considered revolutionary, but her manner of presentation set new standards in scientific communication. Carson the scientist wrote for the general public, for the average citizen, and for the nonscientist who wanted and needed to know a great deal more about his and her own environment.

An example of Carson's departure from a strict scientific format can be seen in the opening passages of her book. She begins with a description of a fictional town where "there was a strange stillness." The stillness came about because of the absence of birds and their songs. The result was a silent spring. But what had happened to the birds? According to Carson, there was a death in the normal cycle of nature. The cause of that ecological death came with the use of pesticides.

In the introductory chapter, titled "A Fable for Tomorrow," the author foretells the sudden and unexplained deaths of children:

> Then a strange blight crept over the area and everything began to change . . . mysterious maladies swept the flocks of chickens; the cattle and sheep sickened and died. Everywhere was a shadow of death. The farmers spoke of much illness among their families. . . . There was a strange stillness. The birds, for example—where had they gone? . . . . It was a spring without voices. On the farms the hens brooded, but no chicks hatched.[4]

Carson then reveals the cause of this strange, silent spring; "no witchcraft, no enemy action had silenced the rebirth of new life in this stricken world. The people had done it themselves."[5]

The death of this fictional town, according to Carson, was unmistakeably the result of the careless and excessive use of chemicals in the environment. She writes:

> The most alarming of all man's assaults upon the environment is the contamination of air, earth, rivers, and sea with dangerous and even lethal materials. This pollution is for the most part irrecoverable; the chain of evil it initiates not only in the world that must support life but in living tissues is for the most part irreversible. In this now universal contamination of the environment, chemicals are the sinis-

ter and little recognized partners of radiation in changing
the very nature of the world—the very nature of its life.[6]

Rachel Carson wrote *Silent Spring* barely 20 years
after the introduction of over 200 chemicals created for the
purpose of killing insects, weeds, and rodents. Carson
startled America and, in fact, the world with the first
rebuttal of the impact of pesticides on the ecology. She was
the first scientist to make the public aware of the inter-
woven system of nature, the *ecosystem*, and the effects that
chemicals can have upon that living, physical community.
An ecosystem (*eco*, from the Greek root *oikos*, meaning
house) is defined as a community of animals and plants
and their interrelated environment. In the new decade of
the 1960s, ecology and the ecosystem became an active
part of our language and thought. Thus, Carson was
responsible for the public's awakening:

> These sprays, dusts and aerosols are now applied almost
> universally to farms, gardens, forests, and homes—non-
> selective chemicals that have the power to kill every insect,
> the "good" and the "bad," to still the song of birds and the
> leaping of fish in streams, to coat the leaves with a deadly
> film, and to linger in the soil—all this though the intended
> target may be only a few weeds or insects. Can anyone
> believe it is possible to lay down such a barrage of poisons
> on the surface of the earth without making it unfit for all
> life?[7]

During the 20 years preceding the publication of Carson's book, our society had basked in the glory of an
improved mode of life due to the introduction of chemicals. Our affluent society with its expanding economy
produced unlimited quantities of attractive food, such as
worm-free, shiny red apples and tomatoes. Suddenly the
bubble burst. Were these ominous warnings voiced by

Rachel Carson at all true? Were we in fact poisoning ourselves and our planet?

Rachel Carson became the central figure in the eye of
a powerful storm. While her message stirred concern and
anxiety in some quarters, it led to open resistance and
hostility in others. According to one of her biographers,
Carol B. Gartner,

> the most reprehensible were the personal attacks . . . at-
> tempts to impugn her credentials as a scientist, and accusa-
> tions of inaccuracy, fanaticism, emotionalism and insuffi-
> cient documentation. Charges were noisy and widespread
> but unproven. Many of the critics had never even read the
> book they misrepresented.[8]

A second biographer, Paul Brooks, refers to the man-
made attempts to discredit her and even sabotage the
book. In Brooks's telling biography, *The House of Life*,[9] the
author points out the efforts made by a major chemical
company to pressure Carson's publisher to withdraw
publication of *Silent Spring*. Their attempts were unsuc-
cessful.

The problem of pesticide poisoning was ripe for discussion. In 1962, the year of the book's publication, Carson
met with citizens of Long Island who were suing the
government to protest aerial spraying of DDT to control
gypsy moths over their heavily populated suburban areas.
Meanwhile, across the Hudson River, their actions drew
the wrath of a New Jersey commissioner of agriculture.
"We are confronted," he was reported to have said, "with a
vociferous misinformed group of nature-balancing, or-
ganic-gardening, bird-loving unreasonable citizens."[10]

Such interests as bird loving and organic gardening
used pejoratively sounds humorous to our ears today; but
in the world of the 1950s and early 1960s, a period of

enormous citizen complacency, and the era which econo-
mist John Kenneth Galbraith had labelled, the "affluent
society," such acts of protest and rebellion by citizens was
comparable to heresy. Moreover, criticisms of Rachel Carson's views were coming much closer to home than the
irascible, antiorganic-gardening commissioner of agriculture.

A special publication of *Silent Spring*, undertaken as a
public service project by Consumers Union, a consumer
advisory organization, printed a disclaimer in the fore-
word to their edition. This disclaimer was actually written
by the director of the sponsoring Consumers Union (CU),
Dexter Masters. Hardly a vote of confidence for Carson's
labors, CU Director Masters wrote:

> Consumers Union cannot and does not endorse every
> point Miss Carson makes. She is a respected researcher, a
> trained biologist, and an extremely competent hand at
> expressing her views. . . . From the wide range of her
> factual material she has proceeded to some conclusions
> with respect to dangers to human health which seem to
> CU's medical advisors extreme.[11]

In fact, Rachel Carson's work, viewed with the more
knowing eye of the 1980s, was hardly extreme. It was a
solid piece of work, well integrated with many examples
of the "unseen" danger that chemicals could present to the
environment and, in turn, to our health and well-being.
But Carson was introducing a concept of nature more
familiar to primitive societies than to our nascent, high-
tech world. The emergence of scientific and technological
advances in America had left us virtually unaware of the
delicate balance between nature and the life it supports. In
fact, if anything, we had come to believe that technology
could control and alter any force in nature, and all for the

better. After all, had we not witnessed in that same era the harnessing of the atom, the awesome birth of atomic energy as a clean, limitless energy supply, and the growth of nuclear power? Carson's new technologically oriented audience no longer feared or respected the forces of nature, since they had finally found the scientific means to control those forces, or so they believed. For Carson's ideas to be fully absorbed, her readers would require an education about the ecosystem before they could comprehend the dangers that pesticides posed to that system. For example, water, she writes,

> must be thought of in terms of the chains of life it supports—from the small-as-dust green cells of the drifting plant plankton, through the minute water fleas to the fishes that strain plankton from the water and in turn are eaten by other fishes and by birds, minks or raccoons—in an endless cyclic transfer of materials from life to life. . . . Can we suppose that poisons we introduce into water will not also enter into these cycles of nature?[12]

As these extraordinary ideas spread across the United States and later overseas, phrases like "ecological balance" and "food chain" entered our vocabulary. Carson was practical in her suggestions for correcting these problems as well. In retrospect, we realize that Carson suggested that we employ a technique we considered new in the 1980s: this concept, now known as integrated pest management, advocates employing a variety of measures available to us to control agricultural problems. Those measures include the use of nonchemical biological controls, as well as the selective use of chemicals as needed. In Carson's book, this alternative is described as "the other road."

THE OTHER ROAD

"The Other Road," Rachel Carson's final chapter, offers detailed examples of the progress of entomological research. The experiments at the time attempted to solve the age-old agricultural problems of pest control by offering alternatives to pesticide use. These alternatives included experiments in sterilizing male insects so that they would not reproduce new generations of undesirable pests. Insect sterilization had actually been proposed a quarter of a century earlier, in the 1930s, by the chief of the United States Department of Agriculture's Entomology Research Branch, Dr. Edward Knipling. Though he had met with some bureaucratic inertia, Dr. Knipling and his associates persisted in their research until finally, in 1954, scientists were able to demonstrate the effectiveness of rearing and releasing sterilized male screwworm flys into the environment. The screwworm fly larvae had been a major insect enemy of livestock in the South, because they live in open wounds on livestock and introduce a bacterial infection which eventually causes death. The screwworm fly was reportedly eradicated from the Dutch island of Curacao in the Caribbean after the application of Dr. Knipling's theory. Carson comments:

> Beginning in August 1954, screw-worms reared and sterilized in an Agriculture Department laboratory in Florida were flown to Curacao and released from airplanes at the rate of about 400 per square mile per week. Almost at once the number of egg masses deposited on experimental goats began to decrease, as did their fertility.[13]

The alternative methods of "turning insects against themselves," as Carson suggested, could be accomplished without the heavy spraying of chemical pesticides, although some chemicals might be used to accomplish this purpose in more restricted doses. But this early research in insect and plant control took a back seat to the major advances in pesticide production. According to the burgeoning and lucrative chemical industry, pesticide manufacturers, farmers, food sellers and government agencies, pesticides would catapult the United States into becoming the world's premier agricultural producer. Not to be overlooked was that widespread pesticide use led to the growth of a very healthy U.S. chemical industry. If anything, this powerful new industry had no incentives whatsoever to encourage the research of alternative methods to pesticides. After all, there was not a large capital return on the rearing of sterile insects.

Though Rachel Carson was not alone in pressing for alternatives to widespread pesticide use, she was the most effective writer in reaching the general public. Consequently, she inspired many citizens to rethink the current pesticide mentality and participate in groups to solve the inherent problems. This power to rouse the public no doubt explained the vitriolic response to her work in certain quarters. The fact that she was a woman made her position even more difficult. A year after the publication of *Silent Spring*, Betty Friedan published her own "earth-shaker," *The Feminine Mystique*,[14] espousing the idea that a woman could participate in and contribute to the serious issues of society. That kind of thinking was anathema in the industrially expanding but socially regressive era of the 1950s and 1960s when Carson published her book. Repeatedly Rachel Carson was referred to in reviews as, "that hysterical woman." This sexist dismissal of her work happened often. Even at the time of her death in 1964, *Time Magazine* continued to attack her work and ideas even in her obituary. In a bitterly written obituary, the magazine declared that, "despite her scientific training, she rejected facts that weakened her case, while using any material regardless of authenticity, that seemed to support her thesis."[15]

Is there any validity to their criticisms? We do know that her thesis was well documented with material, even though very little information on the deleterious effects of pesticides was collected or even published at that time. Therefore, much of her information came from interviews with scientists and naturalists who were the first to make observations in the field, a procedure which is hardly unscientific. These observations were made by specialists in the natural sciences, by those trained to study nuances in nature and who had no subjective ax to grind. They were by and large not funded by the chemical industry.

Independent of her efforts, support for Carson's ideas can be found in the early reports of the British Nature Conservancy. This Conservancy was created under the Royal Charter in Britain in 1949 as part of the overall British effort to control land-use practices following the post-World War II boom. One year after the inception of the Nature Conservancy, in 1950, the first complaints were received from local villagers concerning damage caused by chemical sprays to the flora of roadside verges.[16] Ten years later Carson too described the destruction of roadside wildflowers in the United States caused by the indiscriminate use of herbicides.

In Great Britain the large-scale use of chemical pesticides provided one of the first indicators that a chemical revolution was taking place in the countryside during the 1950s and 1960s. One of the problems with regulating this new chemical revolution was the lack of primary information. Chemical technology had not existed long enough for scientists to understand fully its long-term effects. Complaints from citizens or local naturalists appeared to emanate from isolated sources, rather than establishment scientists. At times, descriptions of chemical effects sounded more impressionistic than factual. After all, an opponent could argue, what did it matter if the roadside verges and wildflowers that bedecked the rural countryside in Great Britain or the United States were burnt and brown from the careless spraying of chemicals. There was an aesthetic consideration, of course, but surely not a threat to human health. The fatal flaw in this argument was a failure to recognize the role that wildflowers played as indicators of the new, destructive power of chemicals, and the consequences of its indiscriminate use. But at the time, most information about the effects of chemical uses was fragmented at best.

When Rachel Carson was collecting research for her thesis, there was a paucity of material available to her. No one had previously published national data on the subject or even funded studies to collect that data. The chemical industry, agricultural community, and the government all viewed the use of pesticides as a modern panacea. They would hardly have set about to fund research that would prove otherwise. If any research on the subject of chemicals existed at all, it would have focused on the development of more powerful agents, rather than trying to determine the possible hazardous effects on the environment and public health.

Even to the present time, there are scientists, particularly toxicologists, who speak persuasively about the difficulty—even impossibility—of ever detemining the certain hazards of pesticides or other chemicals. For example, one toxicologist, Manfred Kroger, writing in 1984 stated:

> The term "trans-science" has been coined to describe wisdom that cannot be achieved through scientific methodology, questions which cannot be answered by science. For example, in toxicology we wrestle with scientific uncertainly, namely the absence of risk. It doesn't matter whether it's safety of food additives, pesticide residues in the food we eat, or pollutants in our water, science is inadequate to bring such proof simply because to get the answers we need would be impractically expensive.[17]

Does this statement mean that the absence of documentation on the safety or risks of pesticide residue in the food we eat or pollutants in our water is due to the impractically high costs of conducting such research? What makes this research so difficult and costly, we wonder, more costly than new chemical research, to deserve the label of "impractically expensive"? Risk assessment research deals with the mathematical probabilities of some danger or illness occurring as a result of certain actions or events, and the effects that such dangers may have on the public. This is a matter of computing probability statistics and surely cannot be prohibitively expensive. To research in depth every chemical substance alone and in combination is impossibly expensive, but some attempt has to be made anyway. Some scientists are suggesting that risk or no-risk is certainly not the issue at hand. Rather, scientists face the nearly impossible task of determining how much and how often a substance can be used before it can be considered a "risk." That is very much like asking, how much poison can we actually ingest before it causes our death? Others would ask, but why drink poison in the first place? In the case of pesticides, they would be told in reply, because you have no choice. But is that really the case?

To encourage the powers that be to conduct research and to inhibit the wanton use of poisonous pesticides, the public has discovered its own weapon: the boycott. Public boycotts have occurred infrequently in the past 40 years; yet, when they have occurred, the results have been highly dramatic. They have proved effective when the public is alerted to a potentially dangerous residue in its food.

THE BOYCOTT OF AMINOTRIAZOLE

During the 1950s, the mood in this country was one of great citizen complacency. Of course, there were some lapses, but they were rare occurrences.

When the weed killer aminotriazole (trade name, Amitrole) was distributed as a trial sample by the FDA to the various cranberry-growing states of New Jersey, Massachusetts, Wisconsin, and Oregon in 1959, it was viewed as a possible boon to the endless problems of weed control. In the decades before the establishment of the Environmental Protection Agency in the early 1970s, the federal government's control of the introduction of new chemical substances was negligible. It was entirely normal for the FDA to hand out samples of new chemicals for tryouts despite the lack of previous testing. Hence, the sample of Amitrole was offered to farmers. In some cranberry-growing states, as in New Jersey, the commissioner of agriculture warned the farmers against using the untested substance, but other states apparently did not restrict its use; this created a problem. Since cranberry crops are pooled, there is no way of knowing at the marketplace which berries came from Massachusetts, or which came from New Jersey. Just before Thanksgiving of 1959, however, the American public was told through the news services that this year's cranberries were contaminated with a chemical weed killer, Amitrole, known to cause cancer in animals. One grower recalled in a later interview that after the story was carried in the newspapers, "you couldn't sell cranberries to your mother." One grower in New Jersey remembers that his older brother, who was the boss on the family farm, in despair had gone outside and buried the unused sample weed killer—to this day they do not know where it lies.[18] The cranberry-growing industry was negatively affected for a solid three years before growers were to return to full production again. Some of the smaller growers couldn't ride the storm and went out of business. While all this happened before the publication of *Silent Spring*, we can assume that the public knew little about pesticides at the time, but they did know about cancer.

Was the weed killer Amitrole removed from the market because of consumer buying patterns, or was the public short of memory and lulled into forgetfullness after only three years? In fact, Amitrole was not removed altogether from agricultural uses, though it was banned from any further use on cranberries. Amitrole became a "restricted-use" pesticide. The public was apparently unaware of its continued use. Once the crisis of cranberries was concluded, the interest in potentially harmful pesticides was not sustained. But what was significant about this early incident of pesticide alarm in response to potential cancer-causing chemicals in a pesticide was that the public, once given the information from the media, reacted forcefully to the presumed offending product. Of course, the effectiveness of this unified public boycott of cranberries was aided by the unique place those brilliant red berries occupied in traditional holiday meals. Eaten primarily, though not exclusively, at Thanksgiving and Christmas dinners, the cranberry was not considered a staple in our diet, as say milk or perhaps apples. These necessary products would have been far more difficult to eliminate from our diet for an extended period of time; a boycott of them might not have been as effective.

In the next decade, however, numerous pesticides used on almost every agricultural product made it impossible to react singularly to any one of them. Moreover, chemically dependent agricultural practices had already become the norm for our society. The next major chemical battle to catch the public's attention erupted over the use of DDT. Again, it was Rachel Carson who introduced the topic to the public.

THE BATTLE AND BANNING OF DDT

Despite very little data on the effects of pesticides on humans, Rachel Carson managed to communicate with scientists throughout the United States and the world to find out as much as possible about the effect that DDT could have on people and the environment. At the time, the most compelling evidence came from DDT's effects on wildlife. Carson was probably the first scientist to synthesize all of the available information into a coherent argument, and the prime target of her work was the insecticide DDT.

The vast majority of insecticides fall into one of three groups of chemicals. One group, which includes DDT, is known as the *chlorinated hydrocarbons*. Chlorinated pesticides, or *organochlorines*, were among the world's most widely distributed synthetic chemicals. They are synthetic because no such compound exists in nature. Organochlorines are a class of chemical compounds produced by the addition of chlorine atoms to hydrocarbons. In addition to DDT, they include aldrin, dieldrin, and chlordane, all used to kill termites until 1984, when an emergency ban was put on their use.[19] Other widely used substances within this group include lindane, toxaphene, endrin, and heptachlor. Another major category of synthetic organic chemicals are the *organophosphates*, including Guthion, parathion, malathion, phosdrin, diazinon, and dimethoate. Chemically they are derivatives of phosphoric or triphosphoric acids, and they vary in their toxicity. Parathion is a highly poisonous insecticide, while malathion, on the other hand, is considered relatively safe and is employed to contain locust and Medfly infestations as well as by home gardeners. The third, smaller category is the *carbamates*, including aldicarb and carbaryl (trade name: Sevin).

The first and earliest developed synthetic pesticides, chlorinated hydrocarbons, included persistent or "hard" pesticides that remain in the soil, air, and water long after their initial application. The toxins are accumulated in the bodies and tissue of plants and animals. The highest concentrations have been found in predators, carnivores, and humans.

DDT is manufactured from chlorine, benzene, and alcohol, which form a DDT molecule by reaction and combination. DDT is within the chlorinated hydrocarbon category since it combines chlorine with carbon and hydrogen. It is almost insoluble in water, about .0012 parts per million, but dissolves in ethyl alcohol, chloroform, ether, and several petroleum products. This pesticide is very effective against most insects since it acts on the nervous system; it is taken in through the cuticle, mainly through sensory neurons where it upsets the functions of the nervous system. The final symptoms are convulsions, ataxia, tremors, and paralysis.

Before the introduction of organochlorines into agricultural uses, in the period prior to World War II, the only pesticides in common use were either derived from plants, such as pyrethrum, nicotine, and derris, or were simply inorganic chemicals, such as arsenate of lead. They were chemicals known to the environment, not synthetics. When the first of the chlorinated hydrocarbons, DDT, was introduced, it seemed to have all of the virtues of an outstanding insecticide because of its toxicity to a wide range of pests.[20] As noted earlier, by the middle of 1944, DDT was standard military issue for curbing the spread of typhus, malaria, and other killer diseases borne by insects. As such, DDT was looked upon as a welcome benefit to public health efforts, particularly because of the military fighting in jungles, and with the influx of civilian survivors arriving from filth-ridden concentration camps in Germany and Eastern Europe. If the frail, starved survivors did not perish immediately from the effects of long-term malnutrition, they would undoubtedly have died from the spread of typhus, and many did.

Rachel Carson did not disavow this "humane" function of DDT. Rather, she pointed out that if it was dusted as a powder on the skin, it was a safe method for disease prevention. It became dangerous only if ingested into the body; its effects were particularly pernicious if ingested with fats.

"Disolved in oil, as it usually is, DDT is definitely toxic," Carson noted.

> When swallowed, DDT is absorbed slowly through the digestive tract, and can also be absorbed through the lungs. Once entered the body, the chemical is stored largely in organs rich in fatty substances because DDT itself is fat-soluble. Relatively large amounts can be deposited in the liver, kidneys, and the fat of the large, protective mesenteries that enfold the intestines, as well as the adrenals, testes, and thyroid.[21]

Whether or not any of this disturbing information was known or even acknowledged by most scientists, the public was certainly unaware of it. Rachel Carson was and still is both praised and criticized for making the problems generated by DDT a public issue. After all, DDT was the insecticide of choice in the United States in response to the demands for greater food production while feeding the battle-scarred countries of Europe and Asia. In the 1940s, it was estimated that one tenth of the annual harvest in the United States was destroyed by insects.[22] Thus, farmers demanded more and more pesticides. At its peak, DDT was sprayed at the rate of at least one million pounds per year. It is not surprising that a multimillion dollar pesticide industry evolved almost overnight in the United States. This country led the world in the production and consumption of pesticides and continues to do so to this present day. Moreover, exports of DDT to other countries even outdistanced its domestic use. Production reached its peak at 188 million pounds of DDT in the 1963 crop year, of which 114 million pounds was exported. Domestically, in 1964 and 1966, over 95 percent of the DDT utilized by farmers was applied to cotton, 46 percent of tobacco acreage was treated with DDT, and significant amounts of

DDT were used on fruits and vegetables. In 1964, 1.9 million pounds were used on .7 million fruit acres; and in that same year, 1.7 million pounds of DDT were used on .7 million acres of vegetables. Crops such as peanuts, corn, wheat, hay, and soybeans were treated primarily with DDT. On a domestic basis, by 1968, only 72 pounds of DDT was used in disease control programs against forest insects, though much of the pesticide was purchased for overseas use in malaria control by the Agency for International Development and the U.N.[23]

Resistance to DDT

Despite this burgeoning pesticide industry with high demands for DDT, the domestic use of the pesticide ultimately dropped more than 60 percent in the decade between 1959 and 1969. This drop in usage was largely the result of two significant changes: one was a growing insect resistance to DDT and other chlorinated hydrocarbons, such as toxaphene; and the second change was the growing concern about DDT's effect on the environment, particularly after publication of *Silent Spring*.

The first cases of DDT resistance among agricultural insects appeared in 1951. Only about six years following its first use, species of fruit moths, codling moths, leafhoppers, caterpillars, mites, aphids, and wireworms had developed immunities to DDT sprays.[24] One logical response to this worrisome development was to accelerate the research and production of more powerful chemicals, as well as locate substitutes, and this in fact was done. There was a shift in the use of phorphorus insecticides between 1964 and 1966. Methyl parathion was now used on cotton to control the cotton bollworm, a major pest of the delta states; but DDT was not eliminated from use.

The environmental concerns over DDT spraying centered on several issues: one, the long-term indestructibility of the chemical once it had entered the ecological chain, and two, the subsequent effects of DDT on wildlife and ultimately on human health. Finally, it was the effect of DDT on wildlife that actually led to its banning in 1973. Whereas Rachel Carson brought the problem to the attention of the public, it wasn't until scientists came up with hard data that the government started to listen.

Broken Eggshells

In the United States, naturalists and wildlife experts working out in the field in the latter 1960s began to notice that populations of brown pelicans, ospreys, bald eagles, and peregrine falcons were declining. They found that DDT had accumulated in the birds' tissues and thinned the shells of their eggs, causing reproductive failure to these predators. About that same time in England, since the introduction of DDT to the environment, there had been sharp dip in the total thickness and calcium carbonate content of shells in the eggs of peregrine falcons, sparrow hawks, and golden eagles. The evidence was building against chlorinated hydrocarbon pesticides and particularly DDT, which was used so extensively. The research that followed these observations focused on calcium metabolism. Since birds have a higher level of lipids (fatty substances) in their plasma, especially at egg-laying time, DDT, which is fat soluble, would tend to concentrate more readily in birds and their eggs. DDT and other chlorinated hydrocarbons interfered with normal calcium metabolism, and this interference resulted in the thinning of eggshells. In California, for example, it was noticed by scientists that brown pelicans would crush their own eggs simply by sitting on them. The shells had become thin and would crack easily, long before the chicks were ready for hatching. As a result, new life was cut off before birth. Strong evidence of DDT presence could be found in the shells of the unborn chicks of those predatory birds highest on the food chain. It was these birds, the eagles, the ospreys, and the pelicans, who ate contaminated insects, smaller birds, and fish and therefore imbibed and retained heavy doses of DDT. With this evidence, DDT was finally proven unfit for further use to many skeptics. The legal battle that ensued involved scientists, lawyers, federal and state governments, industry, and the public. This three-year battle proved to be the first round in a series of U.S. controversies over the use of pesticides in agriculture.

DDT on Trial

A new classification of professionals emerged, known as environmentalists. They were lawyers, naturalists, biologists, geologists, social scientists, and a host of others who had become involved in solving the problems of the environment. Private citizens concerned about their health and surroundings formed groups to save the environment. Environmentalists initially became involved in litigation against the use of DDT in 1966, when a class action was brought against the Suffolk County Mosquito Control Commission in New York (to abate the degradation of Long Island's natural resources). From the start, the battle to ban DDT was fought on a county-by-county or state-by-state basis. It would take four more years before a national policy on DDT was formed and several more before DDT was officially banned from use in the United States.

Battles that were waged on the heavy use of pesticides in our food found new advocates in young lawyers who were educated during the strong social climate of the 1960s. That unique social climate was characterized by an active commitment to society as well as to the individual. Thus, in 1967, a group of professionals, mostly lawyers, known as the Environmental Defense Fund (EDF), incorporated and promptly filed suit against the Michigan Department of Agriculture to prevent a proposed application of dieldrin (a pesticide in the same organochlorine family as DDT). They also filed a suit against several municipalities that were using DDT to control Dutch elm disease.

The EDF was partially successful in their suit by obtaining court orders against the use of DDT for Dutch elm disease in 55 municipalities. The Michigan Department of Natural Resources also was convinced that a stand had to be taken against DDT, and the insecticide was ultimately cancelled for use in the entire state of Michigan. This new advocacy group, the EDF, soon began receiving support from old, established conservation groups in their legal actions. In concert with the EDF, The Izaak Walton League, a naturalist group founded in the nineteenth century and named for the author of the classic book *The Compleat Angler*, requested a ruling in Wisconsin, stating that DDT was a pollutant of the state's waters under the definition of Wisconsin's water pollution control statute. That same year, the National Audubon Society joined the EDF and filed a petition with the Agriculture Department requesting issuance of cancellation notices for all chemical poisons containing DDT. The petition also asked for suspension of the registration of these pesticides pending completion of cancellation proceedings. The Sierra Club, a national environmental and conservation organization formed in the late nineteenth century, also became active in the battle against the use of DDT. New, locally based, environmentally concerned citizen action groups were formed and united their petitions with the old-guard national conservation groups. There was the West Michigan Environmental Action Council and the Citizens Natural Resources Association in Wisconsin—organizations which had not existed before the battle over DDT. This hotly disputed issue drew people and groups together who had never before involved themselves with political crusades. They found themselves fighting to change large-scale practices and laws that had been considered incontrovertible only a few years earlier. Some significant victories spurred them on. A 1969 ruling in Wisconsin proclaimed that DDT and its metabolites (the products of the chemical breakdown of DDT) were environmental pollutants within the definition of Wisconsin laws, since they contaminated the air, land, and waters of the state and presented injury to public health and wildlife.

This legal action on a state-by-state basis might have continued for decades if the EDF had not finally decided to attack the federal government for its lack of initiative in protecting the environment from the hazards of DDT. The advocacy group filed a petition with the U.S. Department of Agriculture (USDA) requesting issuance of cancellation notices for all economic poisons containing DDT and suspension of registration of these pesticides pending completion of cancellation proceedings. In fact, two years earlier, in 1967, the USDA had cancelled registration of

DDT uses to control house flies and roaches, since the pests had become resistant to the insecticide. The USDA had also cancelled registered uses for DDT on the foliage of more than a dozen crops and had requested the National Academy of Sciences to make an intensive study of the impact on the environment of persistent pesticides, including DDT. The EDF, however, was not impressed with federal actions. It viewed the cancellations as unenforceable, since DDT could just as easily be purchased in large quantities for other purposes than what the cancellation stipulated. Apparently the EDF was correct in its criticism, as the old Federal Insecticide, Fungicide and Rodenticide Act (FIFRA) did not allow for suspension of registrations. (The government ultimately agreed with the EDF, and three years later a new pesticide law, the Federal Pesticide Control Act of 1972, was signed.)

Under FIFRA, a cancellation order was issued when there was a finding that a chemical may pose a significant threat to the environment. The purpose of the order was to initiate an administrative review which could have included development of a report by a National Academy of Science advisory committee. This report could be accompanied by a public hearing, litigation, and a final order. A suspension order was only issued if it was determined that a chemical posed such an immediate threat to man and his environment that the suspension had to be enforced during the administrative review process.[25] In other words, it was extremely difficult to cancel the registration of a chemical, and equally as difficult to suspend its use while the lengthy process of investigation was going on. The EDF complained that the continued registration of DDT allowed the manufacturers to search out alternative means of introducing the pesticide, and therefore bypass actions to contain its use.

The EDF was tenacious in returning to the courts for action. After one skirmish, the U.S. Court of Appeals for the District of Columbia required the Agriculture Department to review its decision to refuse suspension of DDT registration. The department responded that "DDT does not pose any imminent hazard to the environment" and therefore saw no need for suspension. Once again the EDF returned to fight the decision in the courts. It was not until June of 1972 that conclusive action was finally taken. William D. Ruckelshaus, Administrator of the new Federal Environmental Protection Agency (formed in 1970 to take over major responsibilities for federal regulation of economic poisons as well as to protect the general environment), ordered a ban effective on December 31, 1972, on almost all remaining domestic uses of the toxic pesticide DDT. A *New York Times* article reporting on the decision at the time concluded that in the end, the risks were not acceptable. The continued use of DDT over the long term, except for limited public health uses, was an unacceptable risk to the environment, and very likely to the health of humans.[26]

After a decade-long struggle, and after almost three years of legal and administrative proceedings including reports by scientific bodies and a seven-month hearing, 125 witnesses, 370 exhibits, and 9,300 pages of testimony, Ruckelshaus announced on June 14, 1972, the final cancellation of all remaining uses of DDT, effective on December 31, 1972. The EPA administrator concluded that the total volume of DDT use in the United States, about 12 million pounds per year, posed an unacceptable risk to humans and the environment.

DDT products had been used to protect over 50 food crops, including almonds, apples, apricots, asparagus, beets, blackberries, blackeyed peas, blueberries, broccoli,

Brussels sprouts, cabbage, cauliflower, celery, cherries, collards, cucumbers, currants, eggplant, kale, mango, melons, mustard greens, potatoes, prunes, pumpkins, strawberries, turnips, and walnuts. It was used to control parasites on beef cattle, goats, sheep, and swine, and for termite protection on seasoned lumber, finished wood products, and buildings—commercial, institutional, and industrial establishments. It pervaded all the nonfood areas in food-processing plants, in restaurants, on flowers and ornamental plants, in commercial plantings, and on lawn and ornamental turf areas. So extensive was the use of this pesticide that U.S. farmers had to be allotted time for instruction in the use of a substitute pesticide.

Although the EPA was the new federal agency on the block, and took most of the credit in the public's mind for the ultimate ban on the pesticide, the Department of Agriculture had taken measures to protect people and their environment from the hazards of DDT some two years before the ban became effective. In 1970, the Agriculture Department had cancelled DDT uses on about 30 crops to avoid illegal residues. But in the public's mind, the EPA was unquestionably the "Good Knight" of federal government agencies in the 1970s. The agency became the protector of our air, water, oceans, land, and food for nearly 10 years; it then fell into disarray in the early part of the 1980s amid scandals of corruption and conflicts of interest. One top administrator was tried and given a prison term for her gross mishandling of the agencies mandate.

In summary then, with the publication of *Silent Spring* in the early 1960s, the intense battle over the use of chemicals in the modern world had only just begun. Within two years of the book's publication, the author, Rachel Carson was dead of cancer. To some, she was a heroine. From other quarters she was severely criticized for ideas, research, thesis and conclusions. Most often, she was called an alarmist. In retrospect, we realize that it was her revaluations concerning the dangers of pesticides that was alarming to the unsuspecting public. The conclusive evidence against DDT would tend to support her original thesis. The biologist's work has not gone unnoticed. DDT was eventually banned from use in the United States. However, to this day it is still used in Central and South America, and in other countries of the world whose food we import.

Rachel Carson's concern was not simply to ban pesticides, but rather to end their indiscriminate and excessive use. Where alternative biological solutions might be employed she felt they should be given equal consideration. But it would take another 20 years, well into the late 1980s, before the question of chemicals—their effect upon the environment, and their cost to our society—would once again occupy center stage. The most compelling and vital arguments against excessive use of chemicals revolved around the question of their effects on human health. New evidence has only recently come to light on this issue. This evidence will be discussed in the succeeding chapters, as we examine the health risk to farm workers and to the millions of consumers of food.

PESTICIDES AND HEALTH
Are There Dangers?

It had taken us some 20 years to recognize that certain pesticides, such as DDT, could be harmful to the continued existence of wildlife; but we still knew very little about its effects upon human health. Slowly, as the evidence began to collect, we learned about environmental changes going on around us—but the connection of these changes to the health and safety of human beings was barely understood. By the time the damaging potential of pesticides like DDT was publicly acknowledged, some irreversible harm had already occurred. These kinds of organochlorine pesticides, the earliest-used synthetic compounds with extraordinary insecticidal properties, were found to persist almost indefinitely in the environment, moving up through the food chain, from plants to animals to humans. Though not the earliest scientist to make this observation (the first records of human fat storage of DDT were published in 1950[1]), Rachel Carson alluded to the proposed risks of DDT. She pointed out that individuals with known exposure to DDT physically store in their tissues an average of 5.3 to 7.4 parts per million (ppm); agricultural workers store an average of 17.1 ppm. Later, in 1967, it was found that occupationally exposed persons store up to 648 ppm.[2] There were, however, conflicting reports on the potential harm that DDT and other chlorinated hydrocarbons would have upon human health. A National Academy of Sciences (NAS) report issued in 1971 stated that "DDT concentration in man varies from one geographical region to another, and in different economic classes with blacks generally having higher residues."[3] It is obvious that agricultural workers suffer the greater exposure. In the cotton belt, at least where DDT was used extensively on the multitudes of pests that plague the cotton plant, the farm workers would have been African-Americans to a greater extent than other ethnic groups. But, the NAS report affirmed that the acute toxicity of the insecticide to mammals was extremely low. DDT was banned from use nevertheless, even though the argument over its potential harm to human health, at least in some quarters, appeared to be lacking in consensus. And though there was a difference of opinion on its problematic effects, that particular pesticide, which was so widely used domestically, was placed under extensive controls outside the United States. In Canada, Denmark, Italy, and Norway, DDT was banned from use even before the United States had taken that final step.

Despite the banning of DDT in the United States in 1972, traces continue to be found in root crops such as potatoes; nearly every American still has traces of DDT in his or her body.[4] The FDA monitoring procedures confirm that residues of cancelled pesticides such as DDT, endrin, and dieldrin, the three most widely used organochlorines introduced in the first stage of pesticide use, could be found in food products long after the cancellation and termination of their use. These three pesticides, for example, were cancelled in the early 1970s. The consequences of various pesticide levels in human fat tissue is difficult to ascertain. While we know that pesticides accumulate, what we don't know is the effect on human health over time. That cause-and-effect relationship has not been studied enough to provide the conclusive evidence. Apparently, once the pesticide was banned in the United States, there was far less interest in determining the long-term effects. In the minds of the environmental groups who had fought against the continuance of DDT, it was a closed case. They could turn their energies toward the many other environmental problems. Much of the health and safety data submitted by chemical companies over a 30-year period needed to be completely reevaluated. Inadequate data, under current scientific standards, existed for most of the major previously registered pesticides; and the data most lacking concerned potential chronic health effects. This lack of information stems from earliest statutes created to control the registration of pesticides for use on food. Registration began as an act in 1910 designed to register pesticides to protect farmers who use pesticides from false marketing claims; the health and environmental safety aspects of pesticide use have only recently been added to this 1910 truth-in-packaging legislation. This explains the gap in data gathering: the regulations never caught up with the vast changes in pesticide use. After all, pesticide use in the early 1900s involved compounds such as sulfur, mercury, and copper. Synthetic organic pesticides were developed and used much later in time, though the legislation did not change sufficiently in focus. The lag in legislation and regulation has left us with an enormous hole in the history of health effects over a four-decade period. What we do know is that many of the cancelled synthetic organic compounds accumulate in fat tissues and have been detected in population samples over several decades.[5] There were reports issued in the early 1970s that DDT was being transferred from the mother to the fetus of mammals, including humans. DDT could be accumulated before birth because of transplacental passage of DDT from a mother's circulating blood through the placenta to the developing fetus.[6] What those effects were over time is simply unknown.

Much like the unborn chicks of brown pelicans whose eggshells had thinned and cracked long before birth as a result of exposure to DDT, the young of our human species, with their fast metabolism and growing bodies, would appear to be at greater risk from pesticide residues on food than perhaps any other member of our society—with the possible exception of farm workers themselves. Those workers live with constant exposure to chemicals. While farm workers are a discrete number in our society, children are ubiquitous. Since our future rests with them, it is maddening to realize that no one really knows for certain the extent of pesticide contamination of our food, nor can we present any evidence of potential danger other than the probability statistics used as a research tool in risk assessment.

RISK ASSESSMENT: DETERMINING THE POTENTIAL DANGERS

Risk assessment is a statistical means of predicting the likelihood of future dangers or events that are catastrophic in nature. The system, which is constructed for each situation separately, consists of variable combinations of factors which might influence the outcome of an event. The idea is to eliminate flagrant risks and seek alternatives to risky activities and, as much as possible, reduce uncertainties. Probabilistic risk assessment often is used as an objective measuring rod for otherwise politically defined problems. Not only is the system of mathematical probabilities, like the roll of the dice, used as a predictor of future, potentially dangerous events, it purports to predict what that resulting damage will look like. For example, risk assessments can provide us with a mathematical estimate of the probabilities of illness and/or death rates per population. Also, the system is used to examine the social costs resulting from the adverse effects of technological advances. The chemical revolution, for example, is what fostered the birth of pesticides. As we have seen already, there have been social costs connected, as well as some benefits fostered by this birth. Therefore, we may estimate that the risk of a given exposure to certain chemicals can, over a lifetime of 70 years, result in 10 deaths or 10 cases of cancer per 100,000 people. This particular system is the means by which the EPA establishes the limits or acceptable levels of risk when the population is exposed to pesticides and other chemicals. Some ratios can exceed what has been established by the EPA to be the acceptable or safe limit. There is no zero ratio, however. There is always some degree of risk to contend with. The idea is to keep the risk as safe or low as possible. Risk assessments can swing in either direction during an argument, and it is not unheard of for opposing risk assessments to be presented at a public controversy. These differences can result from variations posed by the mix of indicators and the inclusion or exclusion of certain variables.

Perhaps the most inaccurate of risk assessment reports was written a few short years before the actual nuclear power plant accident at Three Mile Island. This report, written by a respected scientist, suggested that this type of nuclear power accident had a likelihood of occurring once in a hundred years. Some three years later, the famed accident occurred. Unfortunately, the scientist's report was off by more than a few decades in time, though the analysis was thorough and scientific.[7]

So we may understand that risk assessment analysis necessarily has its share of limitations, primarily because we are predicting the future. We are asserting that a given number of people will become ill and/or die from the effects of technological advances or natural disasters.

INTOLERABLE RISK: PESTICIDES IN OUR CHILDREN'S FOOD

In early 1989, the Natural Resources Defense Council (NRDC) completed the most comprehensive risk assessment analysis ever conducted—either privately or by the government—of children's food consumption and pesticide residues in food. It is a significant study primarily because prior work had never isolated the risk factor with respect to children. This study identified children as the major consumers of certain types of foods and examined the effects of pesticide residues on their health. The findings of the two-year NRDC study, titled "Intolerable Risk: Pesticides in Our Children's Food," have been alarming.

In many ways they have been more frightening than Rachel Carson's *Silent Spring* because of the inclusion of statistical information that was simply not avaliable in 1962.

The report found that the average child receives four times more exposure than any adult to eight widely used carcinogenic, or cancer-causing, pesticides in food. As a result of their exposure to these pesticides alone, as many as 6,200 more children than would have may develop cancer sometime in their lives. Moreover, these eight pesticides are just a fraction of the 66 pesticides that the EPA has identified as potentially carcinogenic that might be found in a child's diet. (With this information, we must recall that we are dealing with probabilities and not historically documented facts.) But the greatest source of cancer risk identified by NRDC comes from apples, apple products, and other foods, such as peanut butter and processed cherries, that may be contaminated with daminozide (trade name: Alar) and its metabolite, UDMH, which is what daminozide breaks down into during processing. In fact, the report goes on to say, average exposure to UDMH may cause one cancer case for every 4,200 children exposed by the age of six alone—which is 240 times the cancer risk that the EPA considers acceptable following a full lifetime of exposure. Among other carcinogens posing risks to children are the fungicide Mancozeb and its metabolite ETU, found in foods like tomato products and apple juice, and Captan, a fungicide used on crops such as strawberries.

Neurotoxic organophosphate pesticides are designed to poison insects' nervous systems and can cause nervous system damage or impair the behavioral system in humans at levels above what the federal government considers safe. According to the report "Intolerable Risk," at least 17 percent of the country's 18 million one-to-five-year-olds are being exposed to neurotoxic organophosphate pesticides through tomatoes and tomato products, green beans, orange juice, and cucumbers, among others.[8]

The focus of this report on the potential health hazards to children is significant for a number of reasons. For one, children consume fruits and vegetables at a much greater proportion of their total diet than adults. Produce makes up about one third of the average child's diet. The typical preschooler's diet is dominated by fruits, which are the foods most likely to be contaminated by pesticide residues. Thus, the NRDC asserts that the EPA, which regulates pesticide exposure through registration, failed to take children's eating patterns into account when it set virtually all current legal limits for pesticides in food. Instead, almost all of these limits were based exclusively on largely outdated estimates of adults' food consumption levels. In setting all the current legal limits, known as "tolerances," for pesticide residues in food, the EPA relied on estimates of adults' food consumption which were developed in the late 1960s. However, the consumption of fresh produce has increased significantly in recent decades. If the federal regulatory agency has underestimated adult produce consumption in the present decade, it has surely underestimated the diet of children. Hence, the NRDC argued that the EPA underestimated the risks of daminozide residues in apple products, especially relating to the health risks to infants and children.

To understand this seeming "catch 22" in pesticide regulation and registration, we must first try to unravel the abstruse history of pesticide regulation in this country.

REGULATING PESTICIDES: A 40-YEAR DEBATE

After the rapid growth of the pesticide industry, Congress, in 1947, enacted the Federal Insecticide, Fungicide, and Rodenticide Act. In Washington shorthand, this act is nearly always referred to by the acronym of FIFRA. The new Act was written to protect farmers from ineffective and dangerous pesticides, but the concern was mostly about ineffective products which might not perform as the company claimed. This product assurance was accomplished through the registration of labels required on all pesticides. The regulatory authority to control pesticide use demands the following requirement: before a pesticide can be marketed, it must be granted a "registration." The decision to issue a registration for any given pesticide is based on a determination of what uses are considered safe and what restrictions may be necessary.

During the late 1940s, FIFRA was under the jurisdiction of the U.S. Department of Agriculture (USDA). When the EPA was created in 1970, FIFRA's authority was transferred to that agency. Then, in 1972, Congress enacted the Federal Environmental Pesticide Control Act. These new amendments provided for greater responsibility with regard to controls on the use of pesticides, including classification of selected pesticides into a restricted-use category. Most importantly, a national monitoring program for pesticide residues was organized. The 1972 Act required a reregistration of older pesticides to ensure that they met the new data requirements. It is this last requirement, the reregistration of older pesticides, that has caused the greatest concern because proof of their unhealthful effects is a lengthy, tortuous, and often near-impossible task.

Newer pesticides are tested prior to registration, while older pesticides already in use for some 40 years now require a series of tests not previously applied to them as part of the reregistration process. And we are speaking of literally thousands of chemicals to be reexamined. In 1989, the GAO (Government Accounting Office) estimated that there are 35,000 pesticide products which must undergo reregistration. Further, the GAO estimated that under the EPA's existing procedures, it would not complete reregistration until the year 2024. But even the most rapid action would not allow the task to be completed much sooner. The long-term animal laboratory studies used in this procedure require up to four years to complete, and then the EPA requires a minimum of one year to review the data, and additional time is needed to implement any appropriate regulatory changes.

It is not too surprising that the EPA, a relatively recent arrival on the Washington scene (1972), had difficulty with its pesticide regulation program. The supporters of more stringent regulations were as critical of the performance of the federal agency as were those urging less government interference in the pesticide field. Later on, this battle over regulation was largely won by the latter group, that is, those who wanted less interference. This was certainly reflective of many governmental policies in the decade of the 1980s. But it is this reregistration process which has left the door open for the continuance of daminozide. Many of the oldest registered pesticides came into existence in the 1950s and 1960s. Under the first 1947 FIFRA, pesticide product registration was permanent. Once registered, there was no need to review the chemical's performance or effects ever again. Then, the updated 1972 amendments made registration renewable on a five-year basis. This process of reregistration had a deadline for completion by 1975 and required a review of all then-registered products to reassess the safety of their continued registration. But the reregistration process completely broke down when the EPA realized that health and safety data submitted over the previous 30 years were inadequate—they were limited in their data base, particularly concerning potential health dangers. It will take many years and research dollars to ultimately ban the use of any previously registered pesticides that have already proven to be carcinogenic or nerve damaging by outside research organizations. It is therefore not surprising that there is a move underway to change the FIFRA regulations, and this movement is spearheaded by the legal branch of the National Resource Defense Council. But they are not alone in this mission. The U.S. Senate Subcommittee on Toxic Substances, Environmental Oversight, Research and Development to the Committee on Environment and Public Works have made similar recommendations in their October, 1989, report.[9]

The current information for registration covers a wide spectrum of health and environmental criteria, such as product chemistry, aquatic toxicity, mutagenicity, and carcinogenicty. This information is evaluated to determine which uses of the pesticide will be allowed for specific pest control at specific application rates. The amount of pesticide residue allowed to be on the treated commodity, called the residue tolerance, is also part of that determination. The EPA sets tolerances defined for pesticide residues in foods; and tolerances define the maximum amount of a pesticide residue that can legally occur on a food or in animal feed. Products with residues exceeding the tolerance are considered to be adulterated and subject to seizure.

Tolerance-setting authorities began in 1954 and were separate from pesticide registration from the start. Until 1970, pesticide residue tolerances were set and enforced by the FDA; that authority was transferred to the EPA in 1970. However, unlike pesticide registrations, which technically have to be periodically reregistered, tolerances are granted for an indefinite period, and there is no explicit statutory procedure for reviewing and revising them on a periodic basis. The crux of the problem lies with the pressure to change the Delaney Clause.

THE DELANEY CLAUSE: AN EXAMPLE OF INCONSISTENCY

The Delaney Clause of the Federal Food, Drug, and Cosmetic Act of 1954 offers the best example of regulatory confusion and inconsistency with regard to protecting the public from cancer-causing products. The House Select Committee to Investigate the Use of Chemicals in Food and Cosmetics, chaired by Congressman Delaney, issued a report in 1952 on the use of chemicals in food products. The report discussed the large number of chemical compounds that cause cancer in animals. The Select Committee recommended that the House enact legislation to control hazardous chemical substances in the nation's food supply. A number of bills were introduced to address this growing concern over pesticides. Congressman Delaney introduced a bill that included a provision regulating carcinogens, known as the Delaney Clause. The FDA objected to the Delaney provision, but Congressman Delaney persisted in his efforts. When FIFRA was amended, there was an unforeseen splintering which resulted in two entirely different procedures for the registration of pesticides. One procedure regulated new chemicals not yet on the market, and the other applied to chemicals already allowed on the market. This resulted in a "grandfathering" of older pesticides that allowed them to persist in use, while newer chemicals were more stringently controlled. Furthermore, this provision of the law, the Delaney Clause, forbids the residues of pesticides that induce cancer in laboratory animals in any processed food, if those residues concentrate above the level allowed on the raw food. The Delaney Clause, however, does not apply to raw foods with no processed form, or to carcinogenic pesticides that do not concentrate in processed foods. Consequently, residues of the same carcinogenic pesticides are not allowed in processed foods where they may concentrate, but they are permitted on certain other fresh and other processed foods. To complicate the situation, the EPA has applied the Delaney Clause only to new pesticides, thereby maintaining registrations for many older pesticides that pose risks acknowledged by the EPA to be greater than those posed by most new chemicals. It is this infamous clause of the Federal Food, Drug, and Cosmetic Act that is the target of independent environmental organizations acting in the interests of change.

Though the EPA does not disagree with the NRDC over concerns for food safety, and particularly the role of daminozide as a possible carcinogen, there are differences in their methods of formulating risk assessment figures. This difference is significant in determining the magnitude of the risk of certain pesticides, such as daminozide, as well as assessing the urgency for taking immediate regulatory action. It is the NRDC's risk assessment formula that leads to a stronger sense of urgency in the question of continued pesticide use. But just what is this new formula, and how does it lead to such alarming predictions?

ESTIMATING RISKS TO OUR HEALTH

The NRDC report "Intolerable Risk" uses data gathered by the FDA about actual pesticide residues found in interstate food shipments. It uses recently developed government data on food consumption by children to calculate risk. Before this report, the residue tolerance assessment system had always been based on theoretical residues calculated from the allowed application rates for the pesticides. This method, using theoretical assessments, is the way the EPA calculates potential risks and determines acceptable uses of any given pesticide. Herein lies the basis for differing conclusions stemming from risk assessment research. Depending upon the indicators used in the mathematical calculations, the assessment of risk can lead to a variety of conclusions, going anywhere from the range of high risk to no observable effect for the exact same product. When the NRDC changed the methodology by working with a different set of figures, their conclusions were far more alarming than any forecast by the EPA.

PESTICIDE RESIDUES: DETERMINING THE TOLERANCES

The EPA has primary responsibility for the registration and regulation of pesticides and for the enforcement of these regulations. In addition, the EPA establishes "tolerances," or maximum allowable limits, for pesticide residues for each crop or animal feed. The tolerance is the maximum amount of pesticide residue in either parts per million (ppm) or milligrams per kilogram (mg/kg) which may be legally present on foods or feed at the time of sale.

Tolerances are determined as a no-observable effect level (NOEL), which does not produce the adverse effects observed at higher-dose levels. In other words, a tolerance level means that there would be no ill effects from the pesticide, though at greater concentrations of the chemical there might certainly be problems. The actual tolerance is derived from data supplied by the registrant, the company producing the chemical. Then the NOEL is reduced by a safety factor (typically 100), because the extrapolation of toxic doses from animals to humans is only approximate. This calculation results in the acceptable daily intake (ADI) in milligrams of pesticide per kilogram of body weight per day. In other words, multiplied by the weight of an average adult, this becomes the maximum permissible intake (MPI) in milligrams per day. Differences in toxicity among infants, pregnant women, the elderly, or other subpopulations is not considered in deriving the MPI or provided for by the use of additional safety factors.

At this point, the potential dietary intake must be derived from studies of consumer dietary patterns. A food factor is the indicator of the amount of each product consumed, on average, per day. The theoretical maximum residue contribution (TMRC) is the proposed tolerance multiplied by the food factor. If tolerances exist for a pesticide on more than one commodity, the TMRC is the total of all proposed and established tolerances times the food factors. The TMRC must be less than the maximum permissible intake (MPI) for that pesticide. Since the entire argument is theoretical in nature, the EPA choses to err on the side of safety by making the TMRC less than the MPI. But there are problems with the tolerance-setting process as it currently functions under the EPA. As we've noted, the food factors used to estimate potential daily intake may significantly underestimate consumption by some subgroups in the U.S. population—children in particular—according to the recent NRDC report. Moreover, the food factors for each commodity are derived from Department of Agriculture survey data either on agricultural production or on household consumption. In using the agricultural production data, total production of each commodity during 1975 is divided by the total U.S. population. Household consumption data, on the other hand, are currently based on a 1965–1966 USDA survey. This clearly outdated information has resulted in estimates of consumption which may be considerably lower than actual figures for many large groups of consumers. An increasing interest in health and diet in this country has led to larger purchases and consumption of fruits and vegetables, certainly in the 1980s. For example, average U.S. consumption of foods such as mushrooms, nectarines, radishes, summer squash, cantaloupes, eggplant, and tangerines is assumed to be less than 7.5 ounces, or approximately one-half pound per person per year. That figure is astonishing low. Even fastfood restaurants now include salads in their menus. While the EPA is currently revising the dietary estimates for subgroups such as children, the elderly, and ethnic groups, the tolerance-setting process does not yet reflect these important differences in eating patterns and the resultant consumption of pesticide residues. It is for this reason that the NRDC study argued that the EPA has underestimated the risks of da-

minozide residues in apple products, especially with regard to the health risks of infants and children.

The groundbreaking effect of the NRDC report is
twofold: one, the fact that they chose to disaggregate
children as an important subgroup in the evaluation of
risks, and two, their development of a risk assessment
formula based on actual pesticide residues found in food,
rather than the traditional EPA formula based on theoretical residues. By using actual residues, certain food is
much higher than average. The NRDC concluded that the
EPA was permitting unreasonable risks. The EPA coun-
tered that their report follows the risk assessment policies
and guidelines established not only for the EPA but gov-
ernmentwide for all other agencies calculating risk. Yet the
recent report "Intolerable Risk" was not the first to call
the EPA to task for their pesticide tolerance assessment
system.

Over the past 10 years there have been a variety of
critiques by congressional committees,[10] the General Accounting Office (GAO),[11] and the National Academy of
Sciences (NAS).[12] These reports have cited incomplete
data files, antiquated assumptions about the intake of
various foods, inadequate monitoring and enforcement of
existing regulatory requirements, inability to routinely
monitor for the majority of pesticides approved for use on
food, and the slow pace of completing the tremendously
large and complex task of modernizing the pesticide regulatory system.

What the NRDC report adds to the debate is not only
a suggestion that a method more sensitive to the risks of
pesticide residues on infants and children is warranted,
but also the claims that, based on actual residues found in
the American food supply, the current regulatory controls

on pesticide use can leave the consumers exposed to an intolerable risk.[13]

PESTICIDES IN FOOD: SIGNIFICANT RISKS FOR CHILDREN

To develop an adequate data base of preschooler exposure to pesticides, NRDC used consumption data from a nationwide food consumption survey of children and adult women conducted in 1985 by the USDA. In addition, they added data on residue levels of 23 pesticides actually measured in types of fruits and vegetables. The data on pesticide residues in produce were derived from analyses of over 12,000 food samples conducted under regulatory programs of the FDA and the EPA.

The principal findings of this analysis lead to some alarming forecasts for the health of future generations of children. Preschoolers, the report asserts, are being exposed to hazardous levels of pesticides in fruits and vegetables. "Between 5,500 and 6,200 . . . of the current population of American preschoolers may eventually get cancer solely as a result of their exposure before six years of age to eight pesticides or metabolites commonly found in fruits and vegetables." To add to the strength of their argument, the NRDC report goes on to state that these estimates are based on scientifically conservative risk assessment procedures. They indicate that more than 50 percent of a person's lifetime cancer risk from exposure to carcinogenic pesticides used on fruit is typically incurred in the first six years of life. It is during this period of rapid growth and development that a preschooler with a swift metabolism consumes greater quantities of food per body weight than children at an older age.

The potent carcinogen unsymmetrical dimethylhydrazine (UDMH), a breakdown product of the pesticide daminozide, is the greatest source of the cancer risk. For children who are heavy consumers of food that may contain UDMH residues, NRDC predicts one additional cancer case for approximately every 1,100 children, a risk 910 times greater than the EPA's acceptable level.

Since the pesticide daminozide is the connecting link to UDMH presence in food, its role as a growth regulator in apples and peanuts has been of no small interest to the consumer. The so-called apple alarm of 1989 brought into center stage all of the actors engaged in the growing public concern about the potential health effects of pesticide residues in fruits and vegetables.

CONCERNING APPLES AND PESTICIDE

In the winter of 1989, the CBS television program "60 Minutes" broadcast a story about the pesticide daminozide, known primarily by the trade name Alar. The television report referred to claims that the pesticide is present in foods disproportionately consumed by infants and children. Moreover, the report continued, the EPA, by its own determination, concluded that daminozide presents an unreasonable risk of cancer to humans, based on results of laboratory animal tests. If this were so, why then had the EPA failed to ban the pesticide from future use? The time lapse between the first record of hazard evidence in 1973 and the EPA's estimated date for an end to daminozide use is a good 20 years, though in all probability daminozide use would be banned in 1991 or 1992. In fact, after an intensive review of daminozide's risks and benefits completed some four years prior to the publication of the critical NRDC report, the EPA announced its plan to ban the use of this chemical because it presented an unreasonable risk to consumers. That same year, 1985, an independent Scientific Advisory Panel determined that the animal studies used to support the ban were flawed, thereby necessitating completion of more reliable studies before Alar could be legally taken off the market.[16] In short, the Scientific Advisory Panel said that none of the present studies were considered suitable for quantitative risk assessment.

In the EPA critique of the NRDC groundbreaking report on the subject, "Intolerable Risks," the federal agency suggested that the data used to produce the NRDC study were flawed. In fact, all of the claims in this newsworthy report on pesticides in our food are based on the same flawed data rejected by the Scientific Advisory Committee in 1985. But the advisory committee's judgment may have been less than objective. It was criticized by members of congress for a possible conflict of interest.

The Flawed Scientific Advisory Panel

The Senate Subcommittee on Toxic Substances has taken a critical view of the membership of the Scientific Advisory Panel, specifically, its members' connections to the chemical industry. Subcomittee members learned prior to the hearing that the Scientific Advisory Panel, which is supposed to render an independent judgment on the safety of chemicals in our food supply, has a significant number of members who are active chemical industry consultants. In fact, most of those on the panel had been paid consultants to the chemical industry or to organizations supported by the industry at the same time that they served on the panel. (Members can accept funds from a particular company in the years before and the year after they review the chemicals they produce, so the issue of ethics is not that confining.) Two Democratic Senators, Joseph I. Lieberman of Connecticut and Harry Reid of Nevada, both of whom serve on the Senate Subcommittee, as reported by the *New York Times*,[14] "said that seven out of the eight members of that panel had served as consultants to the chemical industry at the time they ruled on the use of Alar."[14]

The *Times* went on to report that one of the panel members now with a consulting concern, worked as a consultant to Uniroyal (just five months) after he left the Science Advisory Panel. Another member, a professor of toxicology at the University of California at Davis, appeared as an expert witness against the EPA to defend the pesticide Dinoseb, which is also made by Uniroyal. He had also been a member of the Scientific Advisory Panel, which had advised the EPA not to cancel Dinoseb. The first of these two scientists, acting as a member of the Advisory Panel had recommended that Alar be allowed to remain on the market. Despite these occurrences, the Inspector General of the EPA informed the Senate Subcommittee that the Public Integrity division of the Justice Department had decided not to prosecute these individuals.[15]

Herein we see the difficulties of determining a solid policy based on animal research studies and risk assessments. Not only are there differences in terms of formulas and variables, there are marked disagreements in terms of data, use of data, sufficiency of data, and finally, application. Some of these differences are scientific in nature;

others are clearly political. How can we interpret the assessments made by independent scientific panels whose members are consultants to the chemical industry? And when they determine that the data for the EPA or subsequent NRDC studies are flawed, are these determinations scientific in nature or political? The Senate Subcommittee points to the fact that seven out of the eight members of the Advisory Panel were paid consultants to the chemical industry. Commented Senator Lieberman, "this raises serious questions about whether American consumers are getting the unbiased judgment on these matters which they are entitled to receive."[16] It is the FIFRA law which establishes the Scientific Advisory Panel. The National Institutes of Health and the National Science Foundation are designated to provide the administrator of the EPA with a list of nominees from which to select the panel members. The statute states that members of the panel shall be selected on the basis of their professional qualifications to assess the effects of the impact of pesticides on health and the environment. To insure multidisciplinary representation as much as possible it would be ideal for the panel membership to include representation from the disciplines of toxicology, pathology, environmental biology, and related sciences. The Senate Subcommittee report makes the following observation on this multidisciplinary representation clause: "It is noteworthy that, as of May 15, 1989, there are no public health experts or pediatricians on a panel which is responsible for making recommendations about pesticides that may have significant effects on the health of consumers and particular effects on children."[17] We can see that scientists are not working in a philosophical vacuum. Their professional opinions are shaped by their discipline and the organizations for which they work. Public health experts may view the evidence from a totally different perspective than the toxicologist, for example. Public health experts may ask what is the health effect upon people, and search for statistical evidence to back themselves up. Toxicologists may focus on chemical properties tested in the laboratory, and their work is often very valuable to industry.

A naturalist goes into the field, because that is where the natural world can best be observed—what is happening out there is very important. The thinning and broken eggshells of Peregrine Falcons can only be seen in their natural habitat. The chemical changes in the composition of the shells had to be tested in the laboratory. All of these experts are called scientists; but none of them observes the same phenomena or conducts investigations in the same manner. But that doesn't mean that an interdisciplinary panel cannot reach an agreement on the question of a particular pesticide. Not to be forgotten, for example, is the fact that both the EPA and the NRDC ultimately agreed that the chemical daminozide should be banned from further use. What set them apart was their sense of timing and urgency. After all, there is a lengthy, 25-year history of regulating daminozide, and in that history we will find the key to the mystery of its continued use.

Daminozide (Alar): A Pesticide with a History

First registered in 1963 for use on ornamental crops and in 1968 for use on food, daminozide is now primarily used on apples. Evidence of daminozide's carcinogenic (cancer-causing) activity in animals was discovered in 1977. Of particular concern is a component and degradation product of daminozide: unsymmetrical 1,1-dimethyl-hydrazine (UDMH). Scientists raised the concern that UDMH may cause cancer in animals. The heat process used to make apple juice or apple sauce causes daminozide to break down into the substance (UDMH) which is believed to be an even more potent carcinogen than Alar itself. If that were not enough, UDMH happens to be a component of rocket fuel propellent and as a result, the U.S. Air Force conducted laboratory tests to determine possible risks from occupational exposures. Though evidence of UDMH's carcinogenic activity was first discovered in 1973, it was not until 1980 that the EPA took steps to begin a special review of daminozide. How could they wait so long?

Commenting on this special review of daminozide, James V. Aidala, research specialist with the Library of Congress in an "Apple Alarm" report for Congress, notes:

> a Special Review is an intensified review of the pesticide's risks and benefits in the light of new data suggesting levels of risk not considered at the time of the pesticide's original registration by the EPA. However, after private negotiations with the manufacturer, Uniroyal Chemical Company, EPA decided against beginning the more intense review of the pesticide's risks and benefits. After litigation initiated by NRDC concerning such private meetings between EPA and registrants of a number of suspect pesticides, EPA reopened the questions of further reviewing daminozide.[18]

In July of 1984, the EPA finally began a special review of the disputed chemical, only to be pulled back by the FIFRA Scientific Advisory Panel, which concluded that the evidence presented was not sufficient to initiate a review of the chemical. This was an odd sort of catch 22 at work. Aidala explained that "as daminozide was an already-registered product, the EPA needed to meet the burden of proof that the pesticide posed an unreasonable risk, and the studies were not sufficiently reliable to meet this evidence threshold."[19]

Apparently, the legislation made it very difficult to ban an already existing pesticide that had been in use for a good many years. In contrast, for a new pesticide the prospective registrant is the one who has to bear the burden of proof that the intended use will *not* present an unreasonable risk. Thus, in the light of the Scientific Advisory Panel comments, the EPA in 1986 announced its decision to permit continued use of daminozide pending submission of new toxicology and residue chemical data.

Three years later, as we entered the first month of 1989, the EPA announced its intention to initiate cancellation proceedings to end the use of daminozide. This was done with the interim results of new toxicology studies. Even so, this lengthy process permits the continued use of daminozide for at least another two or three years, until the final outcome of proceedings and appeals are reached. (Yes, the decision can still enter into long appeal negotiations which can drag on for years.) But the question still arises, why had the EPA failed to take action to ban the pesticide sooner, when the earliest evidence of daminozide's carcinogenic activity in animals was discovered in 1977? The answer lies in the highly complicated evolution of pesticide registration policies that allows for the "grandfathering" of existing pesticides, while new applicants for registration can be eliminated more rapidly. Moreover, the limited amount of data gathering on health effects of chemicals in the past left a series of wide gaps in scientific findings. This problem makes proving the case against a particular pesticide extraordinarily difficult.

In 1982, for example, the House Agriculture Subcommittee on Department Operations, Research and Foreign Agriculture released a Staff Report on EPA's pesticide program, comparing information held by EPA on each active ingredient with the data required under FIFRA law. The findings were startling in terms of their gaps:

- 79 to 84 percent of active ingredients currently registered and used commercially lacked adequate oncogenicity (tumor induction) studies. In other words, there aren't enough studies.
- 90 to 93 percent of active ingredients currently registered lacked adequate mutagenicity studies (genetic changes).
- 60 to 70 percent of active ingredients currently registered lacked adequate teratogenicity studies (birth defects).
- 29 to 47 percent of active ingredients currently registered lacked adequate reproductive studies, such as fertility effects.[20]

This striking gap in significant data on pesticide health effects is the result of a legislative afterthought. In other words, detailed and documented health effects data were not required for submission until the 1960s or even 1970s, all in all some 20 to 30 years following their first registration and use. Therefore, the gap exists as far as time and information. The lack of ongoing data leaves a murky trail for scientists to follow in determining the potential health problems generated by the use of particular chemicals over time. It would appear that with the stringent requirements of the Scientific Advisory Panel to prove the unhealthful effects of an already existing pesticide, it would take many years and research dollars to ultimately ban its use. Moreover, a pesticide already proven to be carcinogenic or nerve damaging by an out-side research organization would lead to conflict. Certainly we need to ask: just what is at stake under the present situation? As the legal machinations of government and their critics move inextricably toward a stronger policy of clean food and water, what dangers exist for the present, and for future, generations? Perhaps the most concrete evidence we have of this danger comes from the NRDC study. The pesticides shown in the Table have been presented in that study as possible human carcinogens (others noted in the study have extremely toxic nervous

A Sample of Early-Registered Pesticides, Their Health Effects, and Uses

Chemical	Yr. reg.	Health effects	Crop uses
Mancozeb	1967	Probable human carcinogen, mutagen; causes birth defects	Apples, onions, potatoes, tomatoes, small grains
Captan	1951	Probable human carcinogen, mutagen	Apples, peaches, almonds
Chlorothalonil	1966	Probable human carcinogen, mutagen	Fruits, vegetables, peanuts
Folpet	early 1950s	Probable human carcinogen	Grapes, apples, melons
Parathion	1948	Possible human carcinogen, mutagen	Citrus, cotton, orchards, vege/ fruits

system effects). It is apparent that most of these chemicals were registered quite early on, in the early 1950s and 1960s, and are therefore a significant ingredient in the U.S. diet that have not been accounted for. Approximately 600 of the 1,200 pesticide active ingredients now in use were registered prior to the 1970–1972 amended study requirements, and data gaps for these pesticides have been a subject of public health concern.[22] In addition, being early registrants in the process, they are no doubt due for reregistration, a process which is not only arduous, but, as we have already seen, fraught with scientific and political complications. The difficulties of changing the current situation can only be looked at from a distance by the average American. Most citizens are not active players in the registration process, though citizen action has certainly contributed more and more to the demand for clean food and water. Moreover, individual states have the power to regulate tolerances of pesticide use within their own domain. For example, following the first Alar alert, the state of Massachusetts, in May, 1986, established much lower tolerances for apple products sold in the state starting on October, 1986. And, as of May, 1986, the state of Maine proposed that no residues of daminozide would be allowed in baby food after October 1, 1986, and no residues would be allowed in heat-processed food packaged after that date. Although these were the only states to take such action, some major retailers and processors of apple products (including the Safeway supermarket chain and the makers of Gerber baby food) have announced that they will no longer buy apples treated with daminozide. But can they always tell? Therefore, despite the complicated federal system of registering and more important, reregistering, pesticides, some states and the food indus-

try seem willing to take action. But it's important to remember that these actions are selective and are not affect-
ing the majority of our nation's food supply. There is still
the health of farmers, farm workers, and their surrounding rural communities to consider. Here too, we find a lack
of accurate data collection throughout the years, and a
highly underreported picture of health hazards. This evidence is even more compelling, for as we learn in the next
section, some important health statistics are available to
present evidence of the effects of pesticide use in our
society.

CHAPTER FOUR

UNCERTAIN HARVEST

Farm Workers and Rural Communities

Agriculture has always been a difficult and arduous way of life, particularly for those who till the soil. Through the centuries agricultural endeavors have been highly labor intensive, requiring the efforts of many hands. Harvesting time in rural communities has traditionally involved the participation of every man, woman, and child in the vicinity. Neighbor helps neighbor pick the crops before they rot in the fields or are devastated by foul weather and ravaging pests. From the earliest pre-Christian era to recent times, harvesttime meant more than the gathering of crops—it was a celebration of the renewal of life itself. The food grown and gathered meant a year of sustenance to be celebrated by dancing and ritual feasts. To envision this season of joyous ritual, we can gaze, as in a time capsule, at the paintings of sixteenth-century rural life as depicted by the Flemish artist Pieter Breughel the Elder (1528–1569). Many of us have seen reproductions of his lusty, vigorously colored paintings of peasant life, portraying people eating, drinking, sleeping, dancing, or making love in the fields, as bundles of harvested grain mark the landscape.

At the turn of the nineteenth century, with the rise of the Industrial Revolution, in Europe and the United States, scores of rural families left the countryside for factory work in the cities. Farming areas with dwindling populations began to import workers from surrounding cities just for the harvesting season, when large numbers of workers were most needed. For example, in the lonely cranberry bogs of the southern New Jersey Pine Barrens, extra pickers were needed to collect the ripe red berries. So, in the late nineteenth century, immigrant Italian fruit pickers and their families were recruited each year in Philadelphia, the nearest large city, by a crew boss and transported to the Pine Barrens as seasonal workers, only to be returned to Philadelphia once the job was done. Some immigrants managed to remain on the farms by marrying into the old Quaker families. In fact, their descendents form some of the most powerful farm owners in southern New Jersey today.[1]

During the changing years of the 1930s and beyond, the situation transformed drastically and many of the small family farms disappeared. Workers for the massive agribusinesses became a separate labor force from the overseeing farmer. They formed a distinctly migratory group that still exists today, whose lives are governed by the seasonal harvesting of crops. Like the fruit pickers in southern New Jersey at the turn of the century, the new migratory workers are mostly immigrants. However, a large majority of these workers today arrive illegally by underground railroads from Mexico or Guatemala. Others are poor African-Americans from the south. But that is where the parallel with the past ends. While the earlier migrant workers came in whole family units, these new workers are primarily solo males, living in substandard and unsanitary housing conditions in the labor camps provided by their owners. Tuberculosis is not uncommon, and drug and alcohol addiction seem to permeate these settlements. Even if they were not exposed to pesticides, which indeed they are, their health would be severely compromised by poor living conditions, as well as the physically demanding and hazardous work.

HAZARDS FOR TODAY'S FARM WORKERS

Coyotes, Ted Conover's recently published anthropological portrait of his journey through the secret world of America's illegal aliens describes many of these problems from the first-hand view of a participant-observer. An anthropologist and author, Conover disguised himself as an illegal alien, walked across the deserts, hid in orange orchards, and waded through the Rio Grande. As part of his research, Conover traveled with a group of Mexican migrant farm workers and worked alongside them in the fields. In one brief section, the author describes the workers' exposure to pesticides while picking fruit in an Arizona orchard:

> The inside of the leather work gloves, the part that touched the fruit, had a dull gray gloss, the color you get when you blacken paper with pencil lead, ... "What's this stuff?" "Pesticidas," he replied. . . . The tree leaves were all covered with a flaky white residue. When you reached a tree, you almost got snowed on by a small shower of dusty white flakes. Some stoical pickers wore bandannas over their mouths, banditlike, in order to avoid breathing the stuff. But most of the others, hot enough already, resigned themselves to inhaling it.[2]

An estimated four to five million people are working in the U.S. agricultural labor force, and approximately half of this number are seasonal workers. Seasonal pickers are the portion of our population most directly in contact with pesticides. Some two million workers then are at the greatest risk from the effects of pesticide use. It is the migrant workers and harvesters who come into direct contact with foliage during the highest pesticide application periods.[3] More than 50 percent of these workers are employed in the state of California, a state which produces almost one-half of all the fruit in the United States.

Back in the 1930s, when John Steinbeck wrote his classic novel *The Grapes of Wrath*, he described the disenfranchised farm workers drifting to California as crop pickers. Because little was known about pesticides at the time, the author seemed to romanticize the new technology in his book. He wrote: "And there are the men of chemistry who spray the trees against pests, who sulphur the grapes. . . . These are great men."[4]

Because California produces the bulk of our nation's fruit crop, and two-thirds of our vegetables along with Idaho, Michigan, Texas, and Washington, it is not surprising that the health problems of farm workers caused by pesticide exposure is of critical concern to the State Department of Health Services and the California Department of Food and Agriculture. The research conducted by scientists from these two agencies offers the clearest portrait we now have on the effects of pesticides on farm workers and on those in surrounding rural communities.

One reason that so much information is now available is the result of the state public health's reporting system. This system requires by law that any physician in California who knows or has reasonable cause to believe that a person is suffering from any disease or condition caused by a pesticide must report such a case by telephone to the local health officer within 24 hours of the initial examination. This public health reporting system has been in effect in California since 1977, and allows for more accurate estimates of the pesticide health problems of workers.

Yet, despite the reporting law for physicians in California (the only state in the nation to require this compliance, with a $250 fine for noncompliance), the State Department of Health Services estimates that only 1% of all pesticide-related illness among farm workers is reported.[5] These health problems have always been difficult to pinpoint, since a number of pesticide exposure incidents can produce vague symptoms such as nausea or a respiratory problem that a physician may not recognize or diagnose as a pesticide-related illness. With more than 13,000 pesticide products registered for use in California, containing more than 800 active ingredients (ingredients in the formula which directly affect the target pest) and more than 1,000 inert ingredients (part of the chemical mix which could cause health problems but does not actively affect the pest),[6] it is not surprising that physicians will not always recognize the symptoms.

Underreporting of illness is not uncommon among poor and indigent workers, who carry no health insurance through their employment. Many try to avoid the cost of seeking medical attention or possible detection as illegal aliens and may not even be aware of the alternatives offered by the U.S. health facilities. With approximately 250 cases among farm workers reported in California each year to the State Department of Food and Agriculture, this suggests an actual total of 25,000 cases among farm workers in California per year. Based on conservative Occupational Safety and Health Administration (OSHA) estimates of approximately 1 million farm workers in the highest risk crops and activities, the national total would be an estimated 80,000 cases annually of pesticide related illnesses.[7]

With such a low reporting rate, the most significant information comes from clinical investigations of fieldworker poisoning. One such investigation, conducted by Dr. Michael O'Malley, Associate Medical Coordinator for the California Department of Food and Agriculture, led to a revised understanding of the health effects of a particular pesticide used on grape crops. Dr. O'Malley reports the following story in the Migrant Health Clinical Supplement (June/July, 1988):

> On August 26, 1987, a 25-year-old member of a thirty-man harvesting crew from a vineyard in Madera, California, became dizzy and disoriented while picking grapes and sought treatment at a nearby hospital. His major complaints on presentation to the hospital emergency room were dizziness, vomiting and mild headache. He was diagnosed as having gastroenteritis, given intravenous fluid, and sent home to rest.[8]

But the illness continued and the man returned to the emergency room exhibiting a very low red blood cell (RBC) cholinesterase. The term *cholinesterase* refers to a chemical in the blood—an enzyme that is necessary for proper nerve functioning. This enzyme is attacked by organophosphorous chemicals, one of the largest groups of pesticides presently used. The patient's heart rate dropped to as low as 33 beats per minute. When he was finally discharged one month later, his heartbeat had increased to a normal 50–60 beats per minute.

This first case was the beginning of an unexpected

series of illnesses between August and September of 1987 among three separate crews of San Joaquin Valley grape harvesters. A total of 78 fieldworkers were involved in the three incidents, with 47 reporting a flu-like illness compatible with cholinesterase poisoning. Since low-level exposure to organophosphate pesticides may produce a variety of nonspecific central nervous system symptoms which also occur with influenza, the actual cause of the illness is often missed. These symptoms include headache, fatigue, drowsiness, insomnia and sleep disturbances, mental confusion, disturbances of concentration and memory, anxiety, and emotional lability.

Organophosphates can inhibit or poison cholinesterase by forming chemical combinations that prevent it from doing its work in the nervous system. Chronic moderate exposure results in a cumulative inhibition of red blood cell activity and plasma enzymes. The appearance of symptoms depends upon the decreased rate in cholinesterase activity rather than the absolute level of cholinesterase activity reached. Workers may exhibit 70 to 80 percent inhibition of cholinesterase enzymes after several weeks of moderate exposure without showing obvious symptoms. Documentation of the poisoning is done by monitoring cholinesterase activity. Cholinesterase determinations have been used since the 1950s to document acute poisonings and patterns of chronic exposure among pesticide applicators, but not crop pickers. This is because farm workers as an occupational group are excluded from OSHA regulations and protection.

Even when workers are medically treated, another problem persists: ascertaining pesticide poisoning through the rate of cholinestase activity. Few fieldworkers have a known baseline of their normal rate with which to measure the change or depression of cholinesterase activity in the body. Still, the test can identify low numbers for cholinesterase activity. As such, it can rule out influenza as the cause of these vague symptoms and more than likely implicate one or several organophosphorous pesticides as the causal factor.[9]

The common factor in all three episodes of the grape harvesting crews onset of illness and evidence of depressed cholinesterase activity was the application of an organophosphate pesticide known as phosalone (trade name Zolone). When fields are sprayed with pesticides shortly before harvesting begins, the workers are kept away for a few days or possibly several weeks, depending upon the recommendations for that specific chemical compound. This is done to assure the workers of a safe reentry into a pesticide-laden field. "Reentry" is the term used to describe this safety precaution. Therefore, it was surprising that in the case of these illnesses, several months had already passed since the last application of Zolone in two of the vineyards. This was well beyond the required one-week reentry interval mandated by the State of California. How then could the pesticide be suspected of causing this illness? Dr. O'Malley concludes that a "careful review of the pesticide application histories and measurements of pesticide residue in the suspect fields as well as elimination of possible nonoccupational risk factors, strongly implicated phosalone as the source of all three illness episodes."[10]

Following the investigation, tight control was placed over all phosalone-treated vineyards in California during the remainder of the 1987 harvest season. (The states have full responsibility for these policies, as the EPA nationally does not conduct field monitoring of agricultural worker exposure except for research purposes.) Moreover, given the information on the health effects of this pesticide, the manufacturer voluntarily relabeled phosalone. What would this accomplish? It may not be applied to grapes and other crops (especially to citrus and apples) which require hand harvesting and extensive contact with foliage.

The close attention paid to the questionable health effects of the pesticide by the California Health Department, as well as the corporate responsibility demonstrated by the chemical manufacturer in relabeling phosalone make this incident a positive example of correcting a pesticide health problem. It shows that vigilance, concern, and sound investigation can work to avoid future health problems.

A different incident, this time concerning orange pickers and severe Omite skin poisoning is far less positive. The problem could not be resolved beyond the borders of the single agricultural producing state of California. By contrast, the grape problem was conceivably solved on a national scale because the manufacturer changed the labeling on phosalone applications.

ORANGE PICKERS AND PESTICIDES

Reentry regulations concerning the protection of fieldworkers from pesticide residues can often be the culprit behind pesticide poisoning. The pesticide Omite-CR (active ingredient propargite) was determined to be responsible for the largest outbreak of pesticide-associated dermatitis (skin rash) in California history.[11] Altogether, 198 orange pickers working for a single citrus packer in California's Tulare County were affected, in some instances with deep burns on the skin. The combined chemical effect of the pesticide along with the excessively hot temperatures in the fields led to the outbreak of a severe rash on exposed portions of skin, such as the neck, chest, hands, and arms. What contributed to this problem was that the seven-day reentry interval for Omite-CR had been falsely designated as safe. This "false" designation of safety was the result of a minor change in the chemical formula. An extra inert ingredient in the pesticide formula was added. This shift in formula is what ultimately led to the skin problems. The change in the ingredients of Omite-CR automatically passed for reregistration by the California Department of Food and Agriculture because the reformulation involved inert, and not active, ingredients.[12]

Herein, we begin to notice the problems engendered by gaps in pesticide registration and reregistration procedures. Active ingredients are examined for registration purposes, but not inert ingredients, even though the latter group can also involve chemicals of a serious nature with potential harmful effects. According to the manufacturer of the Omite-CR formulation, the change involved the addition of inert materials to coat the active granules of propargite. This alteration helped to prevent the active ingredient from making direct contact with the citrus leaves, thereby preventing leaf burn. But, the new coating could slow down the degradation of propargite, which in turn would extend the life of the most poisonous properties in the pesticide formula. By using the "old" reentry regulations, the workers were back in the fields handling the pesticide laden fruit long before it was safe to do so. It is largely a case of a significant regulation gap which resulted in skin irritations amongst the workers handling the leaves. Secondly, there was the combined chemical effect of the heat from the excessively hot days acting on the pesticide. This combination of heat and chemicals caused a severe rash on exposed portions of skin, the neck, chest, hands, and arms. A third compounding effect was the reentry period—the preharvest interval originally designated as safe. The reentry interval for Omite-CR was seven days after spraying.

After exhaustive research and study by the Community Toxicology Unit of the California Department of Health Services, the authors of the report concluded that

> it appeared that Omite-CR (containing 30 percent propargite) was the sole agent responsible for the dermatitis outbreak. Since the product was used within the limits of the label instructions and regulations, it appears that residue degradation was not given adequate consideration in the registration process. It is recommended that an extended reentry interval for the use of Omite-CR on citrus be established based on scientific understanding of its properties. A general recommendation is that pesticide reformulations be given thorough analysis prior to registration with attention focused on changes to inert ingredients as well as active ingredients.[13]

This reentry interval was subsequently lengthened by the state to 42 days. However, the Omite-CR label for other states still allows reentry as soon as the spray has dried, allowing harvesting to begin within one week's time. Therefore, though the California citrus pickers may be protected by the state, workers in the other states, such as Florida, may be vulnerable to severe skin rashes. As we noted, the overseeing of pesticide poisoning problems for workers is left primarily for each individual state to determine, and some states have turned out to be more diligent than others. In California, for example, the total number of pesticide poisoning cases reported annually to the Department of Food and Agriculture in the past five years has fluctuated from approximately 2,100 to 2,500. However, following the field investigation conducted by the County Agricultural Commissioner's staff, the number of cases confirmed as pesticide-related occupational illness has ranged from 1,065 to 1,779. This difference between the number of reported cases and the number of confirmed cases is either due to lack of sufficient data linking pesticide exposure to a specific illness or a result of new information changing the status of a case originally reported as a pesticide-related illness to one that is not.[14] Regardless, it is obvious from these statistics that pesticide exposure can certainly lead to work-related illnesses.

In yet another case recorded in 1980, a crew of 23 farm workers were exposed to pesticide residues in a cauliflower field in the Salinas Valley of California. The workers began their job by tying leaves over the head of the vegetable only six hours after the field had been sprayed with the organophosphate insecticides mevinphos (Phosdrin) and phosphamidon (Dimecron). These compounds are potent cholinesterase inhibitors.

Shortly after beginning work, crew members noticed the onset of blurred vision and eye irritation. They continued to work, and over the next two hours additional symptoms of dizziness, weakness, disorientation, headache, nausea, and vomiting developed. Several workers collapsed, one with a loss of consciousness. The victims were all Hispanic and included six women and ten men. Their ages ranged from 9 to 72 years (three children were younger than 15). One patient was discovered to be five-weeks pregnant at the time of the pesticide exposure. Follow-up by the public health investigators showed the woman to have an uneventful pregnancy and was delivered of an apparently healthly infant. But the problem of pesticide poisoning was once again, as in the orange pickers case, related to premature reentry into the fields following preharvest pesticide spraying.

The problem is correctable—but apparently it takes time for the issue to be recognized and corrected by those in authority. A second tying crew was poisoned in September 1982, with 36 persons affected, and a third cauliflower crew was poisoned in the Salinas Valley in April 1983. Recovery from this type of poisoning is quite slow. While the most severe symptoms recede after 28 days, workers' cholinesterase levels did not reach a plateau until an average of 66 days after exposure, after which most patients continued to report blurred vision, headache, weakness, or anorexia. Since a prolonged illness caused the workers loss of work days and income, many returned long before the poisoning was resolved in their bodies, causing reexposure problems. In response to the problem of residue-induced illness, Monterey County in California has adopted a unique regulation requiring that warning signals be posted in fields where a worker safety reentry interval of 24 hours or greater is in effect.[15] But once again, we see that documentation and correction of the problem of worker poisoning is done only after the fact. Even then, the correction is drafted on a county or possible state level, but not as a national concern. Dr. Molly Joel Coye, formerly with the California Division of Family and Community Medicine, and subsequent Commissioner of Public Health in New Jersey, points out that

> there is no regular biologic monitoring of agriculture worker exposures to pesticides except for periodic cholinesterase tests required of certified pest control operators handling organophosphates and carbamate compounds

on a regular basis in the state of California. There are no
regular examinations or surveys to identify the adverse
health effects of pesticide or pesticide residue exposures.[16]

It is not fieldworkers alone who are at risk from problems.
Any worker who has contact with pesticides directly or
indirectly can become ill. Victims can include pilots of
crop dusters and workers who mix and load the chemicals. The list can become quite long.

WORKERS AND ILLNESS CAUSED BY PESTICIDES

Aerial Applicator.   A pilot can actually become exposed to pesticides while flying the crop dusters. Concentrated, heavy doses of pesticides sprayed from the plane
can also enter the cockpit through normal ventilation systems and inundate the pilot with a powerful dose. For
example, two pilots were killed when their aircrafts col-
lided and crashed while both were applying pesticides.
One pilot was flying a helicopter and spraying methomyl,
and the other pilot was flying a fixed-wing aircraft applying methomyl and mevinphos. Neither crash was connected to mechanical failure, though they could not definitely be linked with pesticide exposure.[17] But significant
reports have been made of pilots' in-flight experiences
while spraying pesticides. They point out that they have
perceived slowed reactions and seemingly poor judgment
in themselves which apparently develops in less than one
hours flight time, and can certainly lead to accidents.

Ground Applicators.   People exposed while applying
pesticide dusts or spray by ground application rigs or
hand-held wands attached to such a rig can also inhale
heavy doses of a chemical, even with precautions taken to

protect themselves. The heavy concentration of pesticides coming from hand-held sprayers can conceivably overwhelm the applicator.

Hand Applicator. People can be easily exposed while applying pesticides by hand-pump, hose-end, or backpack sprayer, duster or aerosol can. Still, all of these methods are in common use. The backpack sprayer is worn on the shoulders like an ordinary backpack, and a pipe is directed from the pack toward the target foliage. Unless extreme care is taken, the worker properly trained, and all the equipment is in good repair, that worker can be dosed quite heavily with pesticides.

Exposure to Concentrate. People involved in handling pesticide products between the packaging and the end use are also vulnerable to contact with potentially poisonous materials. For example, workers mixing, loading, and applying the pesticide glyphosate accounted for 33 documented cases of topical injuries.

Field Fumigator. People exposed to fumigant in the field while preparing a pesticide application rig, driving, dismantling, or performing routine maintainance on a rig have been known to suffer skin rashes or poisonings from chemicals. This also includes tarping (covering) a field after fumigation or pretreating tree holes with a fumigant to discourage their use by insects as a place to reproduce.

Tarp Fumigator. Exposure to fumigants can occur while placing a tarp, fumigating, ventilating, or removing a tarp for indoor or outdoor fumigations.

Manufacturing/Formulation Plant Worker. People involved in the manufacturing or processing of technical or formulated pesticide products can accidentally inhale or come into direct contact with pesticides.

Mixer/Loader for Pesticide Application. People exposed

while removing a pesticide from its original container into a mix tank, transferring it into an application tank, or operating a tank rig can suffer the consequences of exposure.

Exposure to Residue on Commodities.   People exposed to pesticides on food products while in the channels of trade. There appears to be little information at the moment on the effects of pesticides on wholesale food handlers, or even the greengrocer. This would depend, of course, on how long residues on food remain harmful.

Exposure to Field Residue.   People exposed to pesticide residues while working in a previously treated field. This includes the workers who engage in thinning, cultivating, irrigating, picking, and field packing.

Exposure to Other Residue.   This group of workers includes primarily office personnel exposed following pesticide application in and around the office for termites, cockroaches, and other interior pests.

Coincidental Exposure.   This involves people exposed to an application strength dilution, but not directly involved in a pesticide-handling activity. For example, spray drift, as in communities surrounding pesticide-treated fields.[18]

PESTICIDE-POISONING EFFECTS
AND SYMPTOMS

The ways that pesticides can affect humans and other mammals are commonly referred to as modes of action. The modes of action of a number of pesticides used today are unknown or, in some instances, only partially understood. Although we may not know exactly how a pesticide

poisons the body, some of the signs and symptoms resulting from pesticide poisoning are quite well recognized.[19]
Early detection of the signs and symptoms of poisoning
and immediate, complete removal of the source of exposure could save lives.

Although data collection on health effects is relatively
recent given that these chemicals have been heavily in use
for over 40 years, evidence does suggest that pesticide use
creates several immediate health hazards on the farm.
Growing in evidence is the serious health problems for
farm workers and for the farmers themselves. A 1986
study by the National Cancer Institute found that Kansas
farm workers who were exposed to herbicides for more
than 20 days per year had a six times higher risk of developing non-Hodgkins' lymphomas than nonfarm workers.
And follow-up work in Nebraska found that exposure to
the herbicide 2,4-dichlorophenoxyacetic acid (2,4-D) more
than 20 days per year increased the risk of developing
non-Hodgkins' lymphomas (tumors of the lymph system
resulting from this rare form of cancer). Other studies
have suggested a link between pesticide use and increased incidence of multiple myeloma among farmers.[20]
So in addition to the immediate poisoning and life-threatening effects of exposure to pesticides, cancer is implied if
not altogether proven at this date.

The following information describing the effects and
symptoms of pesticide poisoning has been drawn from
the work of Professor Bert L. Bohmont, Agricultural
Chemicals Coordinator for the College of Agricultural Sciences at Colorado State University.[21]

Hazardous pesticide groups fall into 10 separate categories, and may affect humans or animals in varying
modes and degrees.

Organophosphorous Pesticides

This group of chemicals is the most widely used pesticide in the United States. They came into strong use and acceptance after the banning of DDT and other chlorinated hydrocarbons. Organophosphorous pesticides have been considered less pernicious than chlorinated hydrocarbons because of their negligible persistence in the environment after use. But some organophosphorous pesticides can still be quite toxic. The farm workers who became ill in California, as mentioned before, were exposed to organophosporous pesticides, which depressed their cholinesterase activity (to a lesser degree, carbamate pesticides, mentioned below, can result in similar health effects). This organophosphorous pesticide group, the largest in use today, includes insecticides such as parathion, malathion, phorate, mevinphos, diazinon, and others. Toxicity values range from high toxicity for parathion, to low toxicity in the case of malathion. The pesticides can be absorbed through the skin (dermally), orally, or through the inhalation of vapors.

As noted in the case of the grape pickers in California, the pesticides in this group attack a chemical in the blood, cholinesterase, that is necessary for proper nerve functioning. When the enzyme is poisoned, nerve impulse transmission races out of control because of a build up of acetylcholine at the ends of nerve fibers. Muscle twitchings referred to as tremors or fibrillations then become noticeable. Convulsions or violent muscle actions can result if the tremors intensify.

Additional symptoms and signs of organophosphorous poisoning include: headache, giddiness, nervousness, blurred vision, dizziness, weakness, nausea, cramps, diarrhea, and chest discomfort. Advanced stages of poisoning result in convulsions, loss of bowel control, loss of reflexes, and unconsciousness. It is important that quick action be taken to obtain medical attention. Prompt treatment can still save people in advanced stages of poisoning, even though they may be near death.

From these chilling symptoms of pesticide exposure, we can now begin to understand the importance of the medical reporting system instituted in California. Since many of the early symptoms mirror vague, flu-like manifestations, the seriousness of the poisoning could easily be overlooked in a hospital emergency room. This was certainly illustrated in the grape picker incident when the first worker with illness was released from the hospital under the assumption that he was suffering from gastroenteritis, and not pesticide poisoning. It is disconcerting that California is the only agricultural producing state that has mandated a public health reporting system. If little is known or documented today about worker-related illnesses of this kind, the lack of a national reporting system could be the explanation.

Carbamate Pesticides

This group of pesticides is relatively new and includes such insecticides as aldicarb (Temik), carbaryl (Sevin—used on citrus crops, fruits, nuts, and fodder), and carbofuran (Furadan); herbicides such as cycloate (Ro-Neet), diallate (Avadex); and such fungicides as benomyl (Benlate) and febam (Fermate).

The mode of action of these compounds is very similar to that of the organophosphorus compounds in that they inhibit the enzyme cholinesterase. However, they differ in action from the organophosphorus compounds in that the effect on cholinesterase is brief, since the carbamates are broken down in the body rather rapidly. The quick reversal from a depressed cholinesterase level back to normal is so rapid that unless special precautions are taken, measurements of blood cholinesterase of human beings or other animals exposed to carbamates are likely to be inaccurate and may appear normal. These pesticides can be absorbed through the skin as well as by breathing or swallowing. Finally, the symptoms and signs of carbamate poisoning are essentially the same as those caused by organophosphorus pesticides.

Chlorinated Hydrocarbon Pesticides

These are one of the older categories of organically synthesized compounds. This group of pesticides includes insecticides such as DDT, aldrin (Aldrite), dieldrin (Dieldrite), endrin, and chlordane. Many of these insecticides have been restricted in use or banned altogether, as in the case of DDT.

These compounds act on the central nervous system, but the exact mode of action is not known. We do know that these compounds or their degradation products (the chemical which results when the product is subjected to heat or other conditions and broken down) can be stored in fatty tissues as a result of a single large dose or repeated small doses. Fat storage of these pesticides appears to be virtually inactive and of no immediate consequence. But without the benefit of data collection on the effects of these chemicals, long-term consequences of fat storage cannot be ruled out as a possible contributor to serious health effects over time.

These pesticides can be absorbed by breathing fumes, through the mouth, or from contact on the skin. Symptoms of chlorinated hydrocarbon pesticide poisoning include nervousness, nausea, and diarrhea. Convulsions may result from a large dose. Liver and kidney damage have been demonstrated in laboratory animals when administered in repeated large doses, but these signs have not been demonstrated in humans to date.

Nitrophenol Pesticides

This group represents different formulations in the fungicide, insecticide, and herbicide categories. All of the nitrophenol compounds can be absorbed in toxic amounts either orally or by inhalation. Some formulations can be absorbed through the skin.

Symptoms and signs include fever, sweating, rapid breathing, rapid heartbeat, or unconsciousness due to the speeding up of certain body processes and functions. If death occurs after a major exposure, it probably will occur within 24 to 48 hours. However, if the patient receives adequate medical attention and exposure has not been severe, he or she will probably recover. Nervousness, sweating, unusual thirst, and loss of weight have been observed in chronic or extended cases of poisoning.

Arsenic Pesticides

These compounds are used as insecticides, herbicides, defoliants, and rodenticides. Arsenic is absorbed primarily through oral exposure, but some can be gained through the respiratory route. Arsenic poisons cells in certain body tissues and also affects certain enzymes, thereby slowing down the rate of normal body functions. Signs of arsenic poisoning include stomach pain, vomiting, diarrhea, a severe drop in blood pressure, and ultimately death.

Mercury Pesticides

At one time mercury was used in preparing seed treatment compounds, but because of persistent and toxic properties, many uses of the compound have been banned. Mercury can accumulate in the body and may cause permanent damage to the nervous system.

Signs or symptoms of mercury poisoning may be delayed and first appear as tingling in the fingers, tongue, lips, headache, and shakiness; inability to think clearly, write, speak, or walk may occur. Changes in personality may follow.

Bipyridyliums

This group of herbicides, which includes diquat and paraquat, may be fatal if swallowed and harmful if inhaled or absorbed through the skin. Lung fibrosis may develop if these materials are taken by mouth or inhaled. Prolonged skin contact will cause severe skin irritation. Signs and symptoms of injury may be delayed, but there are no adequate treatments and effects are generally irreversible.

Anticoagulants

Commonly used as rodenticides, these materials are of danger only when taken orally, and large doses would be required to cause human death. Anticoagulants reduce the body's ability to produce blood clots and sometimes damage capillary blood movement in the body.

Botanical Pesticides

These pesticides are manufactured from plant derivatives and vary greatly in chemical structure and toxicity to humans. They range from pyrethrum, one of the least toxic to humans, to strychnine, which is extremely toxic. Of this group, nicotine (Black Leaf 40) is one of the most toxic poisons, and its action is very rapid. If nicotine is absorbed through the skin or taken in through the mouth, the person becomes highly stimulated and excitable. In fatal cases of nicotine poisoning, death is usually rapid, nearly always within one hour and occasionally within five minutes, due to paralysis of respiratory muscles.

Fumigation Materials

Most fumigation materials are highly toxic and extremely dangerous when inhaled. Within this grouping there is methyl bromide (Bromo Gas), which affects the protein molecules in certain cells of the body. The signs and symptoms include severe chemical burns, chemically induced pneumonia, and severe kidney damage. Any of these effects can be fatal. Carbon tetrachloride, another fumigation chemical, affects the nerves and also severely damages the cells in the kidneys and liver. Finally, chloropicrin (Picfume) is also a highly hazardous chemical which is very irritating to the eyes.

HOW DOES ACCIDENTAL PESTICIDE POISONING OCCUR?

There are three major parts of the body through which pesticides can enter the human system and cause damage: through the mouth, the skin, or by inhalation.

Dusts and sprays can enter the mouth during agricultural applications; pesticides can be drunk accidentally from unlabeled or contaminated containers; pesticides can be eaten on contaminated food; they can be taken in through the mouth when a worker siphons liquid concentrates; or, transfer of chemicals to the mouth can occur from contaminated cuffs or hands, or by drinking from a contaminated beverage container.

Another method of exposure is through the skin. For example, a person can become contaminated from accidental spills on clothing or skin; dusts and sprays settling on skin during application; spraying in windy conditions; splash or spray that touches the eyes and skin during pouring and mixing; contact with treated surfaces, as in too-early reentry into treated fields, hand harvesting, thinning, cultivating, and irrigating, as well as insect or pest scouting. Children can be poisoned by playing in pesticide mixing or spill areas, or by handling discarded containers. Workers who maintain or repair contaminated equipment can also be exposed through their skin.

Poisoning can also occur while breathing in dusts, mists, or fumes from pesticides; or smoking during application of chemicals. In this way, inadvertently, smoking supplies become contaminated.

PESTICIDES IN THE GARDEN

In addition to the work force connected to commercial agriculture, pesticides are in full use in the lawns and vegetable gardens of ordinary, suburban citizens. Home garden pesticides are just as potentially harmful to people. This danger is compounded by the fact that the average home gardener has little if any training in the handling and mixing of these powerful chemicals. For example, diazinon, which is used to control many mites and insects on fruits, vegetables, ornamentals and lawns, as well as insects on soils and indoors, is also toxic to bees, fish, waterfowl, and other wildlife, and can disrupt the function of the nervous system in humans.[22] Carbaryl (Sevin dust) used on trees, turf, and flowering plant pests is highly toxic to honeybees and can cause loss of appetite, weight and weakness after chronic exposure in humans. Propoxur is a probable carcinogen and can temporarily disrupt the function of the nervous system while it functions as an insecticide for lawn pests and household pests, such as cockroaches and flies. So, the same potential health problems found in commercially applied pesticides can invade the peaceful suburbs, where homeowners cultivate their patch of green grass and gardens. Much less is known about the health effects to suburbanites, as there is no registration requirement that covers voluntary use by ordinary citizens.

PESTICIDE POISONING: A SERIOUS MATTER

From this information we can see that pesticide poisoning is a very serious matter that needs to be understood in a larger context. Not only do we need to look at the problem of who gets poisoned by pesticides, but we must examine what safeguards exist to protect people from this happening in the first place. Worker protection seems marginal given the risks. Communities in the vicinity of agricultural activities may suffer from pesticide poisoning without ever coming in direct contact with the chemicals involved. One such community in California complained of health effects stemming from cotton defoliant spraying and eventually voiced their concerns to the California Department of Health Services.

The Acute Health Effects of a Community Exposed to Cotton Defoliants

In response to complaints dating back to the early 1970s by residents of California's San Joaquin and Imperial Valleys, the California Department of Health Services conducted an epidemiological study.[23] The concern of the community was centered on the possible health effects due to cotton defoliant spraying. The purpose of the study was to determine whether community exposure to approximately 100 pesticides reported to have been used during the defoliation season were the cause of an increase in the incidence of reported symptoms including: headache, nausea, vomiting, eye, nose, or throat irritation, cough, shortness of breath, wheezing, "hayfever symptoms," or "asthma symptoms." Communities also expressed concern about the strong odor associated with spraying. In spite of this history of complaints dating back to the early 1970s, there were no epidemiological data on symptoms associated with community or occupational exposure to cotton defoliants prior to this study. Because the Health Services epidemiologists' focus of concern had been the organophosphate pesticides DEF and Folex, researchers looked for possible associations between exposure and the symptoms associated with mild organophosphate poisoning: diarrhea, fatigue, anxiety, dizziness, and depression.

Residents defined as exposed to the defoliant were in fact living or working within one mile of a cotton field. This exposed group was found to be at significantly increased risk for six symptoms: fatigue, eye irritation, rhinitis, cough, nausea, and diarrhea.

The results of this study substantiated the earlier reports by community residents. Living or working near a sprayed field is associated with incidence rates from 50% to 120% higher than expected for eye, throat and nasal irritation, "allergy symptoms," "asthma symptoms," wheezing, shortness of breath, nausea, diarrhea, and fatigue. The risk is even greater when pesticide application is done by aircraft or when a strong odor is noticed. This certainly suggested that an odor-producing pesticide was causing the symptoms. These results turned out to be consistent with the original research hypothesis that DEF and Folex, and their odorous degradation product, butyl mercaptan, were the potentially causative agents. The presence of DEF in the air of the three cotton-growing communities was confirmed by air monitoring data. So, this is an example of a problem that was not caused by direct contact with the pesticides. Rather, it was airborne exposure, probably as a result of the aircraft spraying method of distribution.

In summary then, we can begin to see that pesticides have a profound effect upon our society. Selectively, workers who are engaged in agricultural activities, or in the mixing, packaging, or transportation of chemicals can be exposed to the ill effects that pesticides may produce in humans. Unfortunately, there is no national requirement for systematic monitoring of the health or exposure to pesticides of the more than two million farm workers, applicators, harvesters, irrigators, and field hands who work around pesticides. Industrial workers, however, as producers of pesticides, are monitored through the Occupational Safety and Health Act (OSHA). This otherwise thorough piece of legislation excludes most all categories of farmers and farm workers from the examination of health effects stemming from their work environment. What could conceivably be a health problem for urban food handlers is simply an unknown, though it is generally understood that the most serious effects of spraying pesticides is already degraded in the field before picking. But this understanding involves the potential problem of pesticide poisoning but not the long-term possibility of cancer. It is this appalling lack of documentation that relegates the issue of health effects to a gray area, one that leaves the way open for dispute, and for the continued use of pesticides.

Suburban gardeners do not know what risks they

pose to themselves in growing their own tomatoes. Furthermore, residents of rural communities can also suffer the effects of pesticide use simply because it is often sprayed into their own backyards. People living within a few short miles of a field where spraying is intense are likely to suffer short-term if not long-term effects of the inadvertent airborne exposure to such powerful chemicals. The runoffs of chemicals into lakes, rivers, and even wells is also a topic worthy of serious consideration. This subject will be examined in the following chapter, as we continue to trace the path of health and environmental problems associated with heavy pesticide use.

CHAPTER FIVE

Water

Water is our planet's most precious resource. The existence of all living organisms, including human beings, would be impossible without it. To sustain all forms of life we need water, preferably clean water, to hydrate the body and grow our food supply. It would be natural to assume that a resource as essential to life as water would surely receive diligent societal protection. On the contrary, our water is consistently polluted and contaminated. This precious resource has been abused since the dawn of the industrial age.

It seems that the purity of our water has been on a doomsday course for quite some time. While industrial waste can claim much of the credit for the deterioration of our surface waters—lakes, streams, and rivers—it is the groundwater, those underground systems which are actually broad sheets of water in porous sandy layers, that are affected the most by agriculture. Called aquifers, these underground water systems provide us with well water as well as irrigation for crops; and this precious water supply is being depleted and seriously polluted through a variety of sources. While surface waters of lakes and streams are subject to some regulations, as unbelievable as it may seem, groundwater is virtually unprotected by law. These federal laws designed to protect water include the Safe Drinking Water Act, the Clean Water Act, and the Federal Insecticide, Fungicide, and Rodenticide Act (FIFRA).

GROUNDWATER: THE FORGOTTEN CHILD OF NATURE

Groundwater collects into strata known as aquifers. Aquifers are located underground and are formed by water from rainfall and melting snow and ice that percolates slowly through the soil and rocks. One extraordinary example exists in the Pine Barrens of southern New Jersey. This aquifer is an enormous underground reservoir that in volume is 75-feet deep, with a surface of a thousand square miles. Rainfall percolates easily through the loose, sandy soil. This receptive, loose soil is the reason that so much water has accumulated in the aquifer; it is also the reason for its vulnerability to contamination.[1,2] Chemicals can percolate easily into the ground and enter the aquifer, polluting the interconnecting system of waterways that fill up the strata for miles around. Fairly early on, this magnificent reservoir of fresh water was recognized as a valuable resource, and with this realization came the potential for exploitation. In the latter part of the last century, a financier named Joseph Wharton planned to buy up the land in southern New Jersey and convert the aquifer to reservoirs. This cleaner, purer water would become the water supply for the city of Philadelphia, whose existing water supply at that point in the nineteenth century was already considered unfit for drinking. The New Jersey legislature intervened and kept the aquifer from being converted into a water provision. In many other regions of the country, however, residents depend exclusively on groundwater for drinking, bathing, and washing. That same groundwater resource is drawn to excess for agricultural uses as well. It is the main source of irrigation for arid areas where crops would not normally grow, and therefore it is often used beyond the limits of supply. Peaked by incentives from the federal farm policies to force higher yields of particular crops, farmers pursue the growth of commodities that are not suited to the soil and the terrain. Hence, they compensate with the heavy use of water drawn from the groundwater supply, for irrigation does improve the crop yield dramatically. For example, the Commodity Credit Corporation (CCC), a wholly government-owned corporation, was created to stabilize, support, and protect farm income and commodity prices. The irrigation of crops supported by the CCC, mostly of corn and wheat, resulted in an increase in production of more than 8 million acres in the Great Plains between 1954 and 1982. Irrigation generally boosts yields from 40 to 100 percent.[3]

Ultimately the level of water in the underground reservoirs sinks to a dangerously low level, and salinization (salt infusion) takes place. It is in this last phase of depleting the reserve that fresh water becomes salty, or brackish, and we begin to destroy our reservoirs. Underground water can become salted or brackish in two ways. First, in the coastal plains, the beds of strata are inclined toward the sea. If too much fresh water is pumped from the aquifer, salt water from the ocean which has been penetrating the soil will slowly fill up the empty space left by the drainage of fresh water. As we all know, if water is taken out, something else will flow in to take its place. Therefore, fresh water is displaced by salt water. Second, in the land-locked Midwest, though the region is far from either body of ocean or sea, there is the presence of the Colorado plateau, which contains salt beds below the surface. When farmers drain water out of the aquifer, this lowers the water pressure and water from down below flows up to take its place. This water from beneath the aquifer has been buried in salt beds and brings salinization to the freshwater supply. In this manner, fresh water is replaced by salt water which is no longer useful for drinking, bathing, or irrigation purposes. If nothing else, this excessive use of groundwater for irrigation purposes is a terrible waste of a scarce and valuable natural resource.

This practice of draining underground water supplies for massive irrigation of crops is happening all over the United States. California, Kansas, and Nebraska utilize more than two million acres each of declining groundwater; and Texas is responsible for the depletion of more than four million acres.[4] To further confound the problem, much of this land produces crops already in surplus, so depletion of this precious underground water supply is done so without responding to a demand for a particular food source or consideration for our need to conserve valuable resources in nature. The decision to grow crops already in surplus has to do with economic incentives created by national farm policies that subsidize the cultivation of specified crops. Of these targeted crops more than 10 million acres of cotton, corn, grain sorghum, and small grains are produced with water from declining aquifers.

In addition, most irrigated acres receive high levels of fertilizers to boost the yields even further. From synthetic fertilizers particularly, high levels of nitrate have been detected in the groundwater. According to the National

Research Council's 1989 report on "Alternative Agriculture," a survey conducted by the U.S. Geological Survey (USGS) of 1,663 counties showed 474 counties in which 25 percent of the wells tested had nitrate-nitrogen levels in excess of 3 milligrams per liter. In 87 of the 474 counties, at least 25 percent of the sampled wells exceeded the EPA's 10 milligrams per liter interim standard for nitrate in drinking water. There is a possible health concern when there is prolonged exposure to levels that exceed this standard. It can lead to oxygen deficiency in the blood, though reported instances of this condition have been rare.[5]

Most of the aquifers in the United States have shown some degree of contamination from chemicals. Those underground aquifers that provide us with well water are dangerously affected by agricultural chemicals. In 1987, the U.S. Department of Agriculture estimated that nearly half of all counties in the country were vulnerable to groundwater contamination from agricultural chemicals. A partial survey of the United States conducted in 1988 revealed that at least 5,000 public and private drinking water wells were closed in the past decade because of pesticide contamination, and many more wells required treatment. In these counties, nearly 50 million residents rely upon groundwater for their drinking water. In all, the hazards posed by chemicals used in agricultural production go well beyond those created by industrial pollution.

NONPOINT POLLUTION

Nonpoint pollution refers to sources of water contamination not detectable from a single point. This pollution affects groundwater supplies and surface water as well. The primary source of nonpoint pollution comes from agriculture. Today, only 9 percent of stream pollution comes from industry, and fully 65 percent is nonpoint pollution, stemming primarily from agriculture. An illustration of point pollution (as opposed to nonpoint pollution) which is not only detectable, but correctable, involves the use of feeder pipes, which are often used to dispose of industrial waste. These feeders can be detected, and the offending company can be fined for illegal dumping of contaminated wastes into a stream or river. Nonpoint pollution however, is another story. This contamination is the result of chemicals sprayed from the air or on the ground which enter surface and groundwater. The chemicals, many of them pesticides, wash into water systems by the runoff from rain and heavy irrigation. In nonpoint pollution, there is no particular point of origin. Pollutants are often carried from dispersed sources into water channels by rain-induced runoff from streets, strip mines, and agricultural fields. The pollution is ubiquitous; therefore there is no responsible party for whom fines and enforcement can be affixed. At the present time, nonpoint water pollution is simply not correctable.

The materials needed for agriculture make up the single largest nonpoint source of water pollutants. They include sediments, salts, fertilizers, pesticides, and manures. Nonpoint pollutants, according to a recent report by the National Research Council, "account for an estimated 50 percent of all surface water pollution."[6] Salinization (salting) of soils and irrigation water from irrigated agriculture is a growing problem in the arid West. Irrigation draws upon groundwater to nuture the arid soil. That water, having washed over minerals, salts, and pesticides in the fields, is returned back into the groundwater system through its normal seepage patterns. As mentioned before, that process is even more rapid when the soils are sandy and porous and allow the water to soak in quickly. Further damage is done by lowering the water table in the reservoirs and bringing about a brackish, salty water where fresh water had previously been. The inadvertent addition of salts and minerals to fresh water means that it cannot be used for irrigation or as drinking water. Aquifers and the wells that draw upon them become contaminated. Reversing that process is not technically practicable.

Another problem compounding the quality of our groundwater comes from the heavy use of nitrate fertilizers which also leach through the ground into the aquifers. Nitrate from manures and synthetic fertilizers is found in drinking water wells in levels above safety standards in several states. Groundwater contamination from our present agricultural activities are considerable: they result from dissolved salts infiltrating from irrigation return flows, as mentioned before, and from spills and losses in storing, transporting, and mixing agrichemicals. In at least 26 states, some pesticides have found their way into groundwater as a result of normal agricultural practices. In California alone, 22 different pesticides have been detected in groundwater because of normal agricultural practices.[6]

Why should we concern ourselves with this careless and flagrant misuse of an unseen natural resource? How could this possibly affect our lives? The answer to these questions is fairly astonishing. Aquifers, our groundwater reservoirs, contain nearly 50 times the volume of the nation's surface waters and constitute 96 percent of all the fresh water in the United States.[7] It is from this source of water that wells are drawn. Groundwater provides the water that is found for the creation of wells. Groundwater is the source of public drinking water for half of the U.S. population—nearly 117 million people; yet, every state in the Union has experienced some degree of groundwater contamination. Within the last few years over 2,800 wells have been closed in California, 2,600 in Long Island, 700 in Connecticut, 500 in New Jersey, and 250 in Massachusetts, as a result of contamination. Furthermore, over 70 percent of the 951 hazardous waste sites on the Superfund priority list (federal environmental clean-up program) as of July 1987 had groundwater contamination.

Increased use of nitrogen fertilizers and pesticides, particularly herbicides, over the past 40 years has raised the potential for groundwater contamination. Feedlots that concentrate manure production also heighten this risk by concentrating nitrates, which, in turn, are added to the pool of groundwater resources. Many widely used pesticides, such as Aldicarb and Alachlor, have the potential to leach into groundwater during normal agricultural practices, and, much to our surprise, the banned but not absent DDT has been detected in recent well samplings.

HEALTH ADVISORY LEVELS FOR PESTICIDES IN WATER

To begin to determine the extent of the problem stemming from pesticide contamination of water, the EPA formulated a predictive scale called the Proposed Lifetime Health Advisory Level, though levels have not been set for all pesticides in use. The median or exact middle figure used by the EPA is calculated in order to reflect the concentration of positive detections for all confirmed studies on a particular chemical. In other words, the EPA pools all the confirmed information on a particular chemical, and then draws upon the median or midfigure from this pool to come up with a standard. Health advisory levels set for pesticide detections in groundwater due to normal agricultural use is measured in ppb (parts per billion). For example, Aldicarb is an insecticide that is very widely used. It is applied to root crops such as potatoes, and chemical residues simply cannot be removed with washing. Aldicarb is set at a health advisory level of 10 ppb. It is startling to discover that median concentration levels as high as 9 ppb (only one part below the established health advisory level) for this one pesticide have been confirmed in the groundwater of the following states: California, Florida, Maine, North Carolina, New York, Rhode Island, and Wisconsin. Aldicarb, the most acutely toxic pesticide registered by the EPA, and one that is currently in use, has been found in 16 states altogether.[8]

In another case, the herbicide Alachlor was recently banned in Canada. Still used in the United States, Alachlor was classified by the EPA as a probable human carcinogen; yet the chemical is the second most commonly detected pesticide in groundwater. It has been found in 11 states at a median concentration level of 0.90 ppm. The EPA Health Advisory Level for Alachlor was set at 1.5 ppm. A bare 0.15 ppb difference exists between what has been found in the groundwater and what the EPA determined to flag as a health advisory level. For carcinogens, such as Alachlor, the Proposed Lifetime Health Advisory Level is based on the exposure levels that present a one in a million risk of cancer in the exposed population. Those states with the highest median concentration levels of 0.90

ppb of Alachlor in their groundwater are: Connecticut, Florida, Illinois, Iowa, Kansas, Louisiana, Massachusetts, Maine, Nevada, Pennsylvania, and Wisconsin. So, if we understand this issue correctly, even if a group were to stop eating fruits and vegetables altogether from this day forward, they would still be exposed to near-danger levels of pesticides through the necessary act of drinking water drawn from wells. While the public may be able to boycott a particular fruit or vegetable in an effort to avoid health-damaging effects generated by the pesticides used, they are far more helpless in the choice of their water. You cannot boycott water and continue to live in a state of hygiene and hydration. And while in many parts of the country the purchase of bottled water has become a matter of choice, if not an absolute necessity, there is no guarantee that bottled water is free of contaminants.

The potential threat of agricultural contaminants, which are the largest contributor to nonpoint pollution and subsequently to the quality of our groundwater, is just beginning to be understood. The EPA is conducting the very first national study which it plans to complete in 1990. Up to now, groundwater problems have been examined on a state-by-state basis, depending upon the enthusiasm of the state to conduct research on water pollution. The EPA study is the first on a national level. After more than 40 years of using pesticides, we wonder why the question had not been raised much earlier. We do know that groundwater in general has been overlooked as a resource for protection. Like so many of our natural resources, it has been perceived as unlimited in quantity and thus expendable. There is a further disadvantage in the fact that groundwater is not visible. Being underground, it cannot receive the attention afforded a lake or river whose depletion and degradation is detectable and noticeable to the average citizen.

Some highly toxic and polluted lakes can actually burst into spontaneous fires and give off noxious odors. One community in southern New Jersey has been witness to this ominous phenomena, affording it the dubious distinction of being number one on the EPA's Superfund list designated for clean-up. Groundwater cannot attract that notice. But why has the groundwater system become so deteriorated in recent years? Has there been a change over time, or are we only recently becoming aware that a very real problem does exist? The rise in population and development of rural areas into suburban settlements seems to be used as the explanation for some of this ill use.

SUBURBAN GROWTH AND WATER POLLUTION

Suburban development had its beginnings in the expanding society of the post-World War I period, when the crowding of urban slums and the need for construction of moderate-cost housing spurred the idea of planned communities in density-controlled greenbelts. Hence, low-cost land close to rapid transit systems of cities was considered the ideal location for this new development. Much of this land was unused or vacant farmland. The idea was to relieve the urban crowding in cities, yet allow the urban workers to commute into the city. But further development of suburbia was deferred by the Depression of the 1930s and the war effort in World War II.

Post-World War II society saw the return to suburban housing development on a much larger scale. The housing shortage at the time was unprecedented, and returning war veterans had certain financial benefits offered them by the government to allow investment into low-cost suburban housing. Massive housing developments mushroomed over rural farming regions, particularly in the densely populated northeast region of the United States. Levittowns, named for their developer, Levitt, provided us with rows of repetitive little boxes of low-cost houses for literally thousands of young, new families.

The sociologist Herbert Gans studied the social evolution of this new phenomena in the early 1960s. I have authored several studies and a book on The Pine Barrens that centered on the environmental effects generated by this rural development.[9] Undeveloped farmland in central New Jersey, for example, where rural Willingboro had once existed, became the site of a Levittown. New housing can place great pressures on the antiquated plumbing and the existing water supply as well. Large-scale housing developments bring an astonishing rise in population to previously rural, agricultural areas. The more primitive waste septic systems for residential uses were originally designed for low-populated rural areas. As those areas increase in use and population, and where sewer lines are not built, septic tank problems result: septic systems that are too small overflow into the soil and, eventually, get into the groundwater, and septic-tank cleaning fluids can percolate into the groundwater. So the evolving of suburbs from formerly rural regions can bring about some serious problems for the quality of our groundwater.

Added to the pressures from expanding populations in the countryside, undeveloped rural regions can also attract industrial development. Some industries contribute their own source of contamination to streams and groundwater systems. Thus, with the development of rural areas the expansion of land use has begun to eat up much of the open space left in this country as well as adding to serious problems in the water supply. Even agriculture has been expanding into previously untouched land. The enlargement of agriculture has meant the tillage of soil not naturally equipped to grow certain crops. That soil is therefore inundated with heavy irrigation and pesticides in attempts to double the output. The farm subsidy law has become linked to agricultural practices which influence the continued use of pesticides and degrade the quality of our groundwater, though that was certainly not its intent at the inception, nor even at the present time.

FARM SUBSIDY AND WATER

The farm subsidy program is a voluntary social contract between the Department of Agriculture and farmers. The government pays cash to farmers for every bushel they produce in exchange for a promise that they will leave a specified amount of their land fallow. The goal is to control production from year to year and to prevent huge surpluses. In this manner, the cost of some 14 farm products is contained at a price level which has been economically healthy for the American farmer. In the grain belts of the U.S. Midwest and Great Plains, subsidy programs are sacrosanct. Any suggestion of even the slightest change in farm policy would have enormous consequences for the healthy U.S. farm economy. Grain farmers, for example, have come to depend on the subsidies for a substantial portion of their income.

Yet these same support programs have serious consequences for the environment. This government farm policy has been in effect for over 50 years, concurrent with the introduction of pesticides in the late 1930s and early 1940s. Though at the start, farm supports were considered to be only a temporary measure to ease the situation during the Great Depression of the 1930s, they have become accepted as an essential ingredient in American agriculture. The first farm subsidy law was signed in 1933, when the economic depression left a large portion of Americans homeless and hungry. The basic thinking at the time was to provide reasonable food prices for the public and assist the farmer to continue producing food. Both of these goals were and still are considered laudable, though certainly the needs have shifted from feeding the United States in the hard times of the economic depression days.

This time honored and long-standing government policy has also touched off certain unintentional consequences. Subsidized crop production practices, as mentioned before, have led to soil erosion and surface and groundwater pollution. These consequences need to be examined when the federal government considers drafting new farm policy programs. The connection between the two issues, the support of agriculture in the United States, and the protection of the environment, has never actually been drafted into one piece of legislation that acknowledges the interplay of these two elements. Outside environmental organizations, such as the Audubon Society and the Natural Resources Defense Council, are asking for new provisions in the law that would reduce the use of toxic chemicals in farming. Current farm policies unwittingly add to the heavy use of pesticides by encouraging the growth of a few specific crops. As mentioned before, the turning away from diversification will lead to an increase in the pest population, as insects or even weeds have an wide area from which to forage without benefit of natural predators. The result has been more pests to be followed by more and more pesticides in use. Further, the incentives to pursue this course in agriculture are purely economic. Through the enforcement of the Department of Agriculture, U.S. farm policy offers financial incentives to produce certain designated crops. And, on balance, the farm policy offers disincentives to farmers to diversify their crops. This is unfortunate because diversification would help the problem of soil erosion and decrease the need to depend so heavily upon pesticides. Moreover, so great are the incentives to cultivate the specified crops, that farmers use land that is not naturally suited to their growth, even arid land, for example. In these arid regions, massive irrigation becomes the norm. In the long run, this is bound to have an effect upon the water table.

Irrigation is an old and respected use of natural resources for agricultural purposes; but recent uses of irrigation from groundwater sources has been excessive. The overuse of surface waters has also been more noticeable. The Colorado River is an excellent example of these excesses. This river is so intensively used for municipal water and agricultural irrigation that in very dry years there has been virtually no water left in the river as it crosses the Mexican border. As to the concern for pesticides entering and contaminating the water system, this is a very recent area of inquiry. California began to investigate pesticide residues in their well water as late as 1986. The findings, which have been included in the annual report to the state legislature, offer us some disquieting information.[10]

SAMPLING FOR PESTICIDE RESIDUES IN
CALIFORNIA WELL WATER

The analytical results of 43,056 well water samples taken from 2,977 wells in 41 counties of California in 1988 showed pesticide residues detected in 115 wells in 14 counties. Of those 115 wells, 109, or 95 percent, were positive for pesticides no longer registered for use in California. How could this be possible?

The ten chemicals detected were: 1,2-dichloropropane (1,2-D), 1,2-dibromo-3-chloropropane (DBCP), dichlorodiphenyl dichloroethylene (DDE), dichlorodiphenyl trichloroethane (DDT), atrazine, bentazon, chlorthal-dimethyl, simazine, trifluralin, and xylene. Six of these chemicals were from point sources (1,2-D, atrazine, bentazon, chlorthal-dimethyl, trifluralin, and xylene). The remaining four chemicals were from nonpoint sources. An example of the ability of some pesticides to endure in the environment long after their use has been banned is the chemical 1,2-D. This chemical has been prohibited as an active ingredient in pesticides since 1984. Yet, the authors of this report found it in sufficient quantities in California groundwater. Moreover, they also detected three other chemicals—DBCP, DDE, and DDT, all of which are no longer registered for agricultural use in California. It was interesting that of all the pesticides detected, only one of them, simazine, was determined to be present in groundwater as a result of legal agricultural use. Simazine was the only chemical determined to be the result of nonpoint source, agricultural-use contamination. So then how are these pesticides entering the groundwater system?

An understanding of how pesticides enter into the groundwater system through agricultural practices is not at all well known. Still we could speculate that they enter the system through:

- Excessive use of pesticides and the method of their application.
- Irrigation practices which draw pesticides into the ground and surface waters.
- The physical and chemical characterstics of a given pesticide.
- The soil type (some soils are porous, and surface fluids filter right down into aquifers).
- The climate.

While scientists work to gain an understanding of these factors, the regulation of pesticides to prevent residues from entering well water continues to elude lawmakers, for both scientific and political reasons.

Some pesticide presence in groundwater is explainable. DDT, long-ago banned from use in the United States, still persists in the environment. Does this explanation alone justify the finding of DDT in California well water where the chemical is no longer even registered for agricultural use? Probably not. One feasible explanation for DDT's continued presence in the environment is the fact that the chemical is banned as an *active* ingredient in pesticide formulation, but not as an *inert* ingredient. A chemical formula for a pesticide consists of a combination of active and inert ingredients. The active chemicals, as the name suggests, are those that are actively engaged in the destruction of the targeted pest. On the other hand, the inert components are included in the formula to assist in the binding or cohesion of the ingredients for a more effective application; but they in no way are considered to be involved in the lethal effects. Nevertheless, the inert category in no way implies that the ingredients themselves are totally benign. This is an important distinction to observe, because in our present pesticide registration system, only the active ingredients are tested for their effects. The inert ingredients are looked upon as a harmless addition to a recipe, much like egg whites added to a mousse for their stiffening effect. The potential threat is that the formula can contain some powerful unmonitored chemicals as well. This gap in the requirements of pesticide registration accounts for the possible presence of chemicals no longer considered in use. Hence, the banning of pesticides only applies to their role as active ingredients in chemical formulas, leaving the door wide open for environmental destruction despite the legislative efforts of state and federal governments. In fact, the legislative efforts of Congress to address the problem of groundwater contamination have met with only minor successes in what appears to be an altogether uphill battle for control of pesticide contamination.

GROUNDWATER QUALITY: THE FEDERAL PROGRAMS AND CONGRESS

Many of the surface water problems, especially from point sources, are now regulated. Therefore, national attention has now turned to the relationship between the contamination of groundwater and agricultural practices in rural areas. There the groundwater is the source of drinking water for about 95 percent of the residents. Added to the problem of groundwater is the depletion of aquifers for use in irrigated agriculture. Nationwide, 68 percent of all groundwater is withdrawn from aquifers for irrigation purposes. Between fertilizer application, pesticide use, irrigation and tillage practices, the use of septic systems, chemical waste dumping, leaking underground storage tanks, injection of wastes into wells, spills, and waste storage pits and lagoons, groundwater pollution has taken on the dimensions of a national problem.

Agricultural activities can contaminate groundwater through spills and losses in storing, transporting, and mixing agrichemicals, and from nitrate fertilizers and pesticides leaching through the ground into aquifers. As noted before, septic tank problems emerge when septic systems are too small and when septic-tank cleaning fluids percolate into the groundwater.

In its 1986 National Water Summary, The U.S. Government Service reported state estimates of groundwater contamination sources.[11] Included in the summary estimates were 22 million septic systems which discharged up to 1,460 billion gallons of waste to shallow aquifers each year. Contaminants included bacteria, viruses, nitrates, phosphates, and organic compounds such as trichloroethylene. Also included were an additional 16,400 active landfills (municipal and industrial solid wastes); contaminants include pesticides, trace metals, acids, and organic compounds.

An estimated 376,000 irrigation wells, 1,900 animal feedlots, and extensive application of agricultural chemicals contribute to the contamination of groundwater. In fact, the EPA, in its report on the prevention of groundwater contamination, noted that agricultural practices were the major source of groundwater contamination in 41 of the states reporting; in six of those states, agricultural activities were the primary source of contamination, exceeded only by septic tanks and underground storage tanks.[12] Compounding the problem even further is the fact that septic tanks and normal agricultural practices both involve legally released pollutants into the environment, and neither is federally regulated. The extent of the severity of the problem is quite difficult to determine. Groundwater analyses for synthetic organic chemicals such as pesticides are scarce, so there is an absence of hard data on the extent of groundwater contamination.

What is already known about groundwater pesticide contamination should indeed provide the basis for national concern, and efforts in the direction of regulation should be started. In 1988, the EPA published an interim data base on pesticides in the groundwater system. The report noted the presence of 74 pesticides in the groundwater of 38 states due to ordinary agricultural practices. In that same year, a partial survey of states revealed that at least 5,000 public and private drinking water wells were closed in the past decade because of pesticide contamination, and many more wells required purifying treatment. But pollution from agricultural chemicals is much more difficult to control than point sources of pollution. Nonpoint pollution does not lend itself to regulatory solutions. It has traditionally been handled on an individual basis, where the agribusiness voluntarily devises best management practices. The bottom line is that this problem, which is national in scope and proportion, is being dealt with on a voluntary case-by-case, farm-by-farm basis. Just how effective this approach can be in addressing the problem is anyone's guess; but it can hardly be considered a comprehensive means of controlling major water pollution problems in this country.

DO STATE LAWS PROTECT GROUNDWATER?

Every state has laws pertaining to groundwater, but they vary significantly; many focus upon water rights issues rather than protection of water quality. Water rights deals with the question of who owns or has the right to utilize a natural resource. Water quality, on the other hand, pertains to whose authority it is to protect the quality of natural resources from degradation. A state's beaches, wetlands, forests, mountains, and other resources are in continual jeopardy from pollution, erosion, mining, solid and hazardous waste disposal, and wetland dredging and filling, to name a few. A state can maintain statutory constraints designed to protect the environment, but that protection is often incomplete. The protection of groundwater is a good example of this sort of regulatory gap.

The EPA estimates that 17 states have enacted specific groundwater protection statutes, and 38 states and territories have established strategies for protecting groundwater quality. The General Accounting Office reported that 41 states have groundwater quality standards, though no two states have identical standards. The interstate routes of many aquifers would transgress state laws in protecting groundwater. After all, water underground does not restrict its movements according to state boundaries; so if one state legislates sufficient groundwater protection statutes, a neighboring state that may share the same aquifer might not. Several states have been very active in addressing groundwater concerns in recent years, including California, Connecticut, Florida, Iowa, Nebraska, and Wisconsin, and these forward-looking state laws have raised questions about the need for greater federal leadership in groundwater protection. While the intervention of the federal system is a logical approach to national issues, its current regulatory system with respect to groundwater quality protection is a maze of fragmented, disconnected authorities. Why is that?

THE FEDERAL DILEMMA IN CONTROLLING GROUNDWATER POLLUTION

At the federal government level, no less than 16 federal departments and agencies are all involved with administering groundwater programs. The very unhappy result is fragmentation and poor interagency coordination. Three of those 16 agencies have the lion's share of responsibility for protection: the Environmental Protection Agency (EPA), the U.S. Geological Survey (USGS) of the Department of the Interior, and the U.S. Department of Agriculture (USDA).

Congress, has addressed some of the major point sources of groundwater contamination, such as underground storage tanks and waste disposal sites; but the hard-to-control nonpoint sources of contamination, and especially pollution stemming from agricultural practices and facilities, prove resistant to law and policymakers. In recent years Congress has made efforts to strengthen federal groundwater protection authority, but there is hardly a consensus on the subject. In the 99th Congress, amendments to the Safe Drinking Water Act and the Superfund law both included groundwater provisions. In the 100th Congress, amendments to the Clean Water Act increased the Act's emphasis on groundwater protection. But protection is still not in place. As we examine the provisions of these various federal Acts, we can begin to understand why groundwater has remained virtually unprotected in the federal domain of environmental controls.

The Clean Water Act

The Clean Water Act has traditionally been applied to the protection of surface waters. When it was first passed in 1972, it was part of the new wave of environmental laws that followed the formation of the Environmental Protection Agency and reflected the rising consciousness in the American public. As never before, the U.S. public understood the pressing need to have government regulate and protect the threatened environment. The Clean Water Act granted the EPA the authority to protect both surface and ground waters, and it contained several groundwater protection aspects:

- Section 208 required states to plan for areawide waste management, including possible protection of the groundwater.
- Section 303 authorized the EPA to set water-quality standards. (Those standards could be applied to groundwater if one assumed that surface and ground waters were interconnected.)
- Section 402 attempts to control point-source discharges into the water through the issuance of a permit system.

In the next decade, the 1987 amendments to the Clean Water Act placed added emphasis upon groundwater protection.

- Section 319 was a new provision that targeted the management of nonpoint sources of pollution. More important, the new section included financial support for protection activities. The 1987 amendment included a 50 percent grant program to carry out state groundwater protection. Such federal grant programs offer the state governments incentives to conduct certain activities by giving half of the revenues needed to the states. The provision gives a matching of funds to instigate some action. However, by the end of 1989, none of the $400 million authorized for Section 319 had been appropriated. Despite this apparent federal dollar inaction, 37 states had completed nonpoint source assessments, and 24 have submitted final management plans.
- Section 520 gave authorization to the EPA to conduct studies on seven aquifer systems to identify the sources of pollution and to offer possible control measures. This part was authorized at $7 million.[13]

Basically, the additions to the Clean Water Act continued to place emphasis upon state groundwater protection activities and management strategies. The federal participation was written largely as enabling state action through some economic supports and direction. But this has been a standard federal–state relationship with respect to effecting national changes in policy and practices. And it is probably the most practical means of achieving these goals. There is, however, one important exception—the interstate boundaries of groundwater. Yet, each state or region has differing conditions in the economy and in the environment. A workable management plan would have to reflect those unique differences. In fact, the state plans which eventually followed the authorization of Section 319 contained a variety of approaches. They included educational programs aimed at manure management, stormwater runoff control, pesticide use, and residential septic tank maintenance. In the vast cycle of instigating change in this country, our system has placed the federal role in leadership to broadly define problems and offer financial incentives to generate state action. Then the state works to provide educational programs to influence specific change in the private sector.

The Safe Drinking Water Act

The Safe Drinking Water Act was designed to protect public drinking water supplies, and it does have four provisions which address underground sources of drinking water.

Under the amended 1986 Safe Drinking Water Act, the Environmental Protection Agency could (1) set national drinking water standards for allowable levels of contaminants in public drinking water supplies (the 1986 amendments required that 83 standards be set for specified contaminants by 1989); (2) establish regulations for the underground injection of liquid wastes into wells; (3) withhold federal funds to projects, such as highways, wastewater treatment plants, and housing, that may contaminate an aquifer that has been designated a sole source of drinking water; and (4) provide federal grants for elective (meaning on a voluntary basis of participation) wellhead protection and sole-source aquifer protection programs.

These four provisions to protect the groundwater are only in effect when the aquifer is a sole source of drinking water for a particular section of the country. When there are several sources at work, such as a lake or river as well as an aquifer, the aquifer is not subject to those provisions. As we understand the thinking behind this provision of the act, there is not a committment to protecting aquifers—only a subportion of this water supply is designated as essential. How will this all play out?

An aquifer is an underground layer of porous rock and sand containing water into which wells can be sunk. It may not be the sole source of water in certain communities, but as lakes and streams of surface water become contaminated, it may at some future date evolve into the only water left fit to use. At that point, it could be too late to protect the aquifer. Looking into the provisions of the Safe Drinking Water Act, which is considered by many to have the greatest potential for preventing groundwater contamination, we can begin to see how short term our thinking has been in efforts to protect the water supply. Further, this Act has certain limitations which would interfere with its implementation. The Maximum Contaminant Levels (MCLs), for example, which the EPA is empowered to set for public drinking water supplies, have only been set for a few chemical groups. The EPA has argued that data on toxicity of drinking water contaminants have been difficult to obtain; though according to the 1986 amendments, the EPA is mandated to develop at least 83 standards in the ensuing years.[14] Since that time the EPA has progressed toward revising the drinking water regulations but not far enough. It is clear, though, that the fewer the standards with which to document contamination, the less that can be said to circumscribe and eventually eradicate the problems. What we are dealing with in the issue of protecting water supplies, particularly ground-water, is an information and documentation gap which continues to place our entire society in the dark with respect to the environment. This kind of information gap also tends to jeopardize the safety and health of our society.

Toxic Substances Control Act, and Federal Insecticide, Fungicide, and Rodenticide Act

Although these two Acts do not engage in provisions for groundwater protection, their implementation could establish environmental controls which would in turn pro-tect the groundwater. This is possible through their role in screening chemicals prior to their manufacture and use.

CONGRESS AND GROUNDWATER PROTECTION

In recent years, Congress has made efforts to strength-en federal groundwater protection authority, but there is hardly a consensus on the subject. In 1987, Congressman Oberstar of Minnesota introduced the Groundwater Pro-tection Act to amend the federal pesticide law. This bill has been a primary focus of a national groundwater protection lobbying campaign conducted by an independent, public interest group called Clean Water Action. In effect, the proposed legislation would require the EPA to restrict the use of cancer-causing pesticides that contaminate the groundwater. Of the 436 congressional representatives asked to cosponsor the bill, only 258 agreed, and the rest declined. All of the voting representatives from Utah, Nebraska, Kansas, Iowa, Idaho, and Nevada declined to cosponsor the bill. In Arizona, only one congressman, the sole Democrat, voted yes. The large agricultural state of California, with its massive voting power of 45 representatives, generated 30 yes votes, a two-thirds majority supporting the idea of the Groundwater Protection Act. Following the political party system, almost without exception, Republicans tended to vote against the Oberstar amendments, and the Democrats in favor. For example, in California, of the 15 nay-voting representatives, 14 were Republicans. Similar results showed up in New York State, where 30 of the 34 congresspersons voted yes, but 3 of the dissenting 4 nay voters were the Republican representatives.

When it comes to congressional voting trends, we do understand that the legislator votes in the direction of their respective constituents. Though the lines of demarcation between political parties are very much blurred these days, by tradition, the Democratic party favored and represented the trade unions and the working class and espoused certain liberal ideas which might include protecting the environment. The Republican party, on the other side, was to represent the causes of "big business." As noted before, these lines have been crossed in many areas, and in recent months, the environment is every legislators' concern, regardless of political affiliation. Nevertheless, the voting patterns on this particular Groundwater Protection Act reflect two major areas of subdivision—political party and region of the country.

With regional subdivisions, the southern states tended to vote for the amendments, with two thirds of the yes votes coming from Florida, North Carolina, Tennessee, Mississippi, and South Carolina. The southern states have fared poorly with the use of pesticides in recent years. The cotton belt has suffered from the effects of pesticide resistant and voracious insects. In recent years, southern farmers have been receptive to integrated pest management assistance from the Department of Agriculture and are amenable to seeking alternatives to pesticide use. Their support for the groundwater protection measure would incorporate this change in agricultural thinking. In general, the northeastern states supported the measure. Massachusetts voted to cosponsor the amendment without exception. New Jersey offered 13 out of a total of 14 votes for the Oberstar plan. The Midwest, as noted above, appeared less than enthusiastic toward federal measures to control the use of cancer-causing chemicals that contaminate groundwater. Even the progressive state of Minnesota, the home state of Congressman Oberstar, offered only a 50–50 split vote. However, public concern in Minnesota has increased with the release of a recent study by the State Health and Agriculture agencies regarding groundwater contamination. The study surveyed 500 private and public wells for pesticide contamination, finding trace amounts of pesticides in nearly 40 percent of the wells tested.[15]

Opponents of groundwater protection counter that federal intervention is unnecessary and inappropriate. They believe that comprehensive protection is achievable through complete implementation and funding of existing programs. Questions such as Should there be minimal national groundwater safety standards? What groundwater resources should be protected? and Should groundwater continue to be protected on a state statute-by-statute basis? persist in slowing the debate on developing federal controls.

Nevertheless, congressional interest in these unresolved issues has continued. In the 101st Congress, pesticide regulation and agricultural management bills were reintroduced; but for the most part, the country was waiting to see what groundwater protection actions the new Bush Administration would take in using federal authority. Taking authority away from states' management of groundwater contamination appears to be a double-edged sword. States such as California maintain more stringent controls and would prefer to continue monitering the situation. Other states choose the flexibility afforded them by the lack of federal involvement. These states would prefer not to have their pesticide-use program become an issue.

The much criticitized Superfund legislation, which is known in Washington circles by the acronym CERCLA, Comprehensive Environmental Response, Compensation, and Liability Act, gives the federal government the authority to respond to problems generated by the discharge of hazardous substances. One section, 101 (8), defines the environment to include groundwater and surface water supplies. But the key issue for groundwater in the Superfund Act is the question of cleaning up contaminated sites. After all, clean-up of contamination is done to prevent toxic substances from seeping into the groundwater. The concern with the Act comes from a proposed groundwater classification system that would classify some aquifers as more valuable than others. And by this classification system, some aquifers would be designated as more valuable to decontaminate than others. But how would they determine these clean-up priorities by which some aquifers were more valuable than others? Aquifer waters can sometimes interconnect. Contamination at one point can certainly reach other sources of water. Therefore, the question is not easily resolved.

We can begin to see the many-faceted problems generated by the continued use of pesticides in our agricultural sectors. If we continue to practice agriculture as we have for the past 40 years, we can look forward to continued and even greater difficulties; and no one member of the society can expect to elude the effects of contamination. Fortunately, there are avenues of change open to us which could redirect the course of our old chemically dependent society. These changes have to do with the same forces that shifted the mode of agriculture in the first place. Agribusinesses that embraced the use of pesticides gave us farming practices that have begun to devastate the environment at levels still not fully understood or documented. It is possible to change our agricultural habits, and we need to comprehend that we have a choice in determining the future safety of our environment and health. The federal government needs tougher policies to ensure us of clean, unpolluted drinking water as well.

CHAPTER SIX

CHANGING AGRICULTURE
The Reformation

Changing our agricultural practices is the key to repairing the damage perpetrated by the chemically dependent society most of us have known for a lifetime. What exactly does this metamorphosis entail, and how can it be accomplished? An agricultural transformation requires the development of new technologies and practices. The goal is to put an end to the use of pesticides, or if that proves impossible, to at least restrict the application of chemicals considerably. Consumers do not appear to have a role in this transformation, at least in the initial stages; but later on they do play an important part in this agricultural revolution. The public's role has always been to press for a more healthful and environmentally safe food supply. Such an interest may require greater interaction with legislators to press for more effective legislation. The American public may also have to readjust their view of cosmetically attractive food. For example, apples may appear less red and rosy when certain chemicals are withdrawn from use.

Consumers will need to understand and appreciate the value of this kind of trade-off.

The new technologies that will supplant the use of pesticides will stem from the innovations of agricultural scientists. Then, following their research of alternative farming systems, they would need the cooperation of the farming community to make it a reality. Farmers would have to agree to adopt and apply new principles to an entrenched 50-year-old agricultural system. This is not a simple task. The agricultural community must first try to reverse the established, long-standing trend of using chemicals as the sole means of pest control. Changes of this magnitude are not usually welcome, convenient, or even economical, especially to those in our society who would have to make the greatest accommodations—the farmers. Many arguments could be raised against taking action that would diminish the role of pesticides—a possible failure of crops, for one. So convinced is the average farmer of the necessity of chemicals as an essential part of our agricultural system that it would be very difficult to convince him otherwise. He would never consider the option unless some practical alternatives were offered to deal with the crop destruction posed by pests. Insects, weeds, mites, and rodents attack many of our most stable food sources, and we can hardly ask the farmers to produce an abundance of fruits and grains without providing them with some form of help. Just call to mind the saga of the locusts described earlier. Locusts are certainly an extreme example of the ravenous destructive power that can attack the orchards and fields of the farmbelt. But if we simply eliminate the use of chemicals without providing effective alternatives, we have failed to address the root of the problem.

If we are to invoke the wisdom of environmentally concerned policymakers, we must also understand that our chemically dependent society is due for an agricultural revolution. If the term revolution seems too strong a word, call it what you may—an overhaul, a revitalization, a change—but hopefully it will be an improvement.

How can this important transformation take place? After all, Americans are used to a highly abundant and easily available food supply. With few exceptions, our country has never been threatened with famine or serious food shortages of any kind. The Great Depression of the thirties brought hunger to our shores, and World War II instituted some selective rationing, mostly of coffee, sugar, tea, rubber, and petroleum. The poor of this country have known hunger on a more or less chronic basis; but for the most part our highly affluent population has rarely been denied, let alone inconvenienced by the lack of available food. We not only expect this opulence to continue, but we tend to see it as a democratic right. And perhaps it is, for by contrast, Eastern Europe has been burdened for decades with shortages of every conceivable kind.

In Romania, for example, for some 20 years the ruling dictator, Ceausescu, withheld important grains and vegetables from his people. Food was used as a bargaining chip in the international game for the balance of payments. Ceausescu starved his people to pay back his hard currency debt to the West. It was the dictator's means of establishing autonomy so that sanctions from the West would count for very little. Thus, the abundant food that Romania produced was traded on the international market, and virtually nothing was available for domestic purposes. Romania has long been blessed with rich soil and generous yields, yet the people were always deprived of this bounty. When this dictator's government toppled abruptly, it was partly the denial of food as well as the brutally repressive system that led to his downfall. If this fallen leader had studied his history more carefully, he would have noticed that the French Revolution in the eighteenth century was precipitated by similar circumstances. This is not to say that Americans will "storm the Bastille" should a certain vegetable or fruit suddenly become unavailable in the marketplace or, worse yet, if the cost of food escalates enormously. Indeed, our unique history in recent times has shown that the issue which is most likely to inspire collective action from the public is an environmental health concern.

While Americans have come to enjoy and probably take for granted the richest food supply in the world, they are not too happy about learning that this same food poses serious dangers to their health. In response to these threats, the public tends to boycott wherever boycotting is possible. In a consumer society, this form of action is a possible and often very effective way of instigating change. Clearly, we understand that in Eastern Europe the citizens had no products to boycott in the marketplace—in fact there was no marketplace with which to interact at all. In a noncapitalist dictatorship, the unpopular leader does not have to worry about pleasing his constituency for fear of being voted out of office. Therefore, he becomes unresponsive to the public's needs. A democratic system, on the other hand, encourages leaders to pay attention to consumer interests, at least part of the time. But at the same time it is crucial for consumers to know how to make their voices heard. One effective way for consumers to make their demands known is for them to be cognizant of the problems and to understand in general terms what solutions are possible to employ. More specifically, ruling out the use of pesticides requires us to have an alternative solution; the public needs to know more about what alternatives are currently available.

THE COSTS TO SOCIETY

When we assess the historical developments and progress of the past 50 years in America, we see that the agricultural revolution which started with the chemical age has caused some unintended and undesirable consequences. Legislative and regulatory controls of pesticides didn't arise until the awakening of our environmental consciousness in the 1970s. Only then could legislators begin to address some of these problems. After all, they needed to catch up with the neglect of the preceding three decades. For those 30 years, from the 1940s to the 1970s, there were virtually no regulations; hence, chemicals were used indiscriminately. Chemicals were intentionally added to promote the growth of food, accidentally added to water, and dispersed within the environment as a whole without anyone's knowing or dreaming of their possible effects. Thus not many scientists were employed to document their impact.

Two monolithic U.S. industries viewed the chemical revolution as altogether beneficial to themselves and, in turn, to the country, since they helped produce a good and bountiful food supply: the chemical industry itself, and the expanding world of agribusiness. Still, despite their optimism, the indirect costs linked with pesticide use quietly began mounting. First came the expected materials and labor costs for pesticide application. And soon, other unanticipated costs began to add up. New chemicals had to be introduced and applied because new breeds of pests kept building up resistances to the old pesticides. There were also large and unexpected costs from human exposure to pesticides. Health problems caused by possible or probable onset of cancer were not without importance, though documentation of these effects as a result of exposure to agricultural chemicals has just begun to surface. The illness and poisoning of farmers and farm workers presented another cost to society that continues today. The increasing costs of pesticides is still another consideration when we recognize that pesticides have become an accepted and often mindless ritual of food production. Our environment is being polluted. The potential danger from the runoffs of chemicals into the soil and water eventually harms the wildlife, the fish, and pollutes an essential, life-giving water supply. And lastly, there are the ever increasing expenditures by the government to enact and enforce pesticide regulations.

These costs are so overwhelming that Cornell University Professor David Pimentel and several of his colleagues added up the indirect costs of pesticides in America today and estimated them to be nearly $1 billion dollars per year. But the most damaging effects in a purely agricultural context is the loss of beneficial, natural enemies of undesirable pests. As an example, pesticides used to control the highly destructive cotton boll weevil inadvertently also manages to kill the natural enemies of the equally destructive cotton bollworm. So while the boll weevil may be temporarily curtailed by pesticides, the bollworm is enemy-free to wreak its own damage on cotton crops. Even malathion, one of the most potentially harmless pesticides, when used for diminishing the damage of insects has been known to kill off a harmless bystander— the honey bee. This "bee kill" happens accidentally when fields are sprayed; yet the aftereffect means that an essential pollinator of flowers and plants has been eliminated. These unintended consequences are part of what Professor Pimentel has referred to as the indirect costs of pesticides.

Indirect costs are defined as any losses due to effects of pesticides on nontarget organisms. The total estimated environmental and social costs of pesticides in the United States adds up to $839,000,000. The following breakdown[1] is a good indication of where the costs are most deleterious:

1.	Human pesticide poisonings	$184,000,000
2.	Animal and pesticide poisonings and contaminated livestock products	12,000,000
3.	Reduced natural enemies and pesticide resistance	287,000,000
4.	Honey bee poisonings and reduced pollination	135,000,000
5.	Losses of crops and trees	70,000,000
6.	Fishery and wildlife losses	11,000,000
7.	Government pesticide pollution controls	140,000,000
		$839,000,000

The largest loss figures stem from the problem of the reduced natural enemies of undesirable insects due to pesticide kills. This estimate also includes the cost of target insects' pesticide resistance, which in turn leads to heavier and repeated applications of chemicals. The second largest figure on estimated environmental and social costs for pesticides in the United States is human pesticide poisoning. This cost includes the necessary medical attention required in severe cases; it does not attempt to calculate the cost of human life when death occurs, nor medical costs due to pesticide-related cancer. The latter concern, as noted before, requires more epidemiological (health history) studies than what is currently available in order to more accurately calculate the societal losses. Pimentel points out that although there is no direct epidemiological evidence that pesticides will cause cancer, the indirect evidence implicates them strongly. For example, 26 pesticides have been found to be carcinogenic in at least one laboratory animal, and some may react to form carcinogens.[2] Pimentel goes on to state that "no one would deny that pesticides have this potential to cause cancer in humans, but whether this potential is actually realized remains to be documented."[3] However, one epidemiological study conducted in 1977 did raise some serious concerns. These scientists reported a significant correlation between the intensity of cotton and vegetable farming and total cancer and lung cancer mortalities in the southeastern United States.[4] The findings of this research deserve further investigation.

But despite that significant gap in estimating the costs of pesticide use to our society, we can already see that there is sufficient evidence to suggest that a transformation in agriculture is in order. Ironically, one solution to the problem created is to turn back the clock—to return to enormously effective pest control techniques introduced into this country in the late 1880s. *Biological control* is the term currently used to describe this particular strategy.

FORGING NEW TECHNOLOGIES:
BIOLOGICAL CONTROLS

The system of biological control employs nature's own predators, parasites, and pathogens (any microorganism or virus that can cause disease) to do the work of pesticides. These natural enemies of undesirable insects and weeds can do the same job without the obvious disadvantages of using chemical controls. The idea is simple to comprehend once we think of nature as a guide, rather than attempting to solve the problem strictly by human mechanics. An appropriate analogy would be that we must learn to use the dog with its herding instincts to protect sheep from predatory wolves. A more lethal solution would be to use chemicals to poison the wolves. However, that approach would not only destroy the wolf population, but also cause harm to other animals who may unintentionally consume the poison as well. It is also conceivable that we can apply these same principles to the protection of crops.

Why haven't we attempted to employ this alternative technology? Is it so untested or unproven that we have simply ignored its possibilities? Not in the least. Research in the development and use of biological controls has been quietly going on in this country and in Canada for a little over 100 years. In fact, in 1989, the United States Department of Agriculture celebrated the one-hundredth anniversary of the use of biological controls against agricultural pests. The Agricultural Research Service within the Department of Agriculture has been diligently working away at nonchemical alternatives for pesticides for quite some time. In fact, the oldest form of biological controls can be observed in a system that is quite familiar to us. The concept is known as crop rotation and was employed as far back in time as medieval Europe.

As the name in crop rotation suggests, the farmer is advised by agronomists to vary the type of crop planted after one or two or even three years until the next cycle. Farmers who use this simple and effective technique allow the naturally occurring, helpful organisms in the soil to work for them and against one or more of their crop enemies. For example, by not growing wheat in the same field more than every second or third year, root parasites are forced into a lengthy dormancy. This inactivity gives the beneficial parasites and predators more time to battle other harmful parasites and sanitize the soil.[5]

In the early 1920s in this country, the African-American botanist and agronomist George Washington Carver taught southern farmers to rotate earth-scarring cotton crops with the cultivation of peanuts. This second crop, undervalued at the time, had the ability to revive the soil. Carver went further in his analysis and suggested a number of effective and profitable uses for the soil-enriching peanut. In 1925, he published a paper suggesting 105 ways of preparing the peanut for public consumption. Hence, the farmer did not need to lose capital while he enriched the soil, and he now had a second market for his crops. It was hoped that this system of crop rotation would also deter some of the devastation wrought by the fearsome cotton-eating boll weevil. This insect, much like the bollworm, destroys the cotton boll or pod and was infamous throughout the southern United States. It even became the subject of some folk songs.

The cotton boll weevil "came a visiting" from Mexico on or about 1892. Not long settled in the U.S. south, the boll weevil became the infamous subject of a number of rural American folk songs. These songs tell the story of the migrating weevil, which devastated cotton and corn crops as it travelled eastward across the United States, all the way to the Atlantic Ocean. One song, called "The Ballit of De Boll Weevil," originated in Texas and Mississippi around 1909. Composed in the oral folk tradition by southern farm workers, the song tells the story of the persistent and damaging insect as a major headache of the cotton and corn farmer:[6]

> De Boll Weevil is a little bug
> F'um Mexico, dey say,
> He come to try dis Texas Soil
> En thought he better stay,
> A-looking for a home,
> Jes a-looking for a home.
>
> Den de Farmer say to de Merchant
> "We's in an awful fix;
> De Boll Weevil's et all de cotton up
> An' lef' us only sticks.
> We's got no home,
> Oh, we's got no home."

The boll weevil is still a source of agricultural woe even after 100 years of living on U.S. soil. The result has been heavy pesticide use to control the insects. In 1980, cotton growers applied pesticides as much as 15 to 18 times before harvest. Then, entomologist Ken Summy, with the Agricultural Research Service, came up with the idea that cotton growers could reduce their pesticide use by 50 percent or more by destroying their plants once the harvest is over. A seven-year U.S. Department of Agriculture study on boll weevils in the Lower Rio Grande Valley of

Texas found that leftover cotton stalks provided great food and protection for the boll weevil during fall and winter months. These leftover stalks offer ideal conditions to produce large populations of weevils to infest the spring cotton crops. By 1984, with the application of stalk destruction, pesticides were applied an average of only 4 or 5 times, instead of the previous 15 to 18 applications. Apparently, eliminating the cotton plants during nongrowth months drastically reduced reproduction, hibernation, and development of weevils in the dried bolls. Over time, scientists believe that the benefits of stalk destruction will be seen in the need for fewer pesticide applications, less risk of groundwater contamination, and lower costs to growers.[7]

Another one of our time-honored methods in the employ of biological controls involves the introduction of natural predators that attack or inhibit a particular pest. For example, we might import the mongoose to kill venomous snakes or devour the less desirable rat population. This is an interesting alternative to pesticides. No chemicals or invasive materials are involved. We need only reshape nature's balance by introducing predators where predators did not exist before. But scientists also need to make certain that these imported creatures do not harm anything else in the environment, especially the crops. Lastly, its important to determine if the new organism can survive in its new setting. The question is clear: why don't we utilize this technology if it is available? There is surely some mystery here that is worth explaining. But first we need to understand the nature of biological controls in order to determine whether the concept really is a workable alternative to the use of pesticides.

BIOLOGICAL CONTROLS: DO WE HAVE
AN ALTERNATIVE TO PESTICIDES?

The birth of biological pest controls began in a citrus grove in 1889 in what is now the city of Los Angeles, California. Citrus trees were overrun with an introduced pest called the cottony-cushion scale. Like many of the agricultural pests farmers contend with, these creatures cross oceans and enter new continents by accident, aboard ships, or hitchhiking in grains and other imported foods. That is precisely what occurred to the fledgling citrus industry in California before the turn of the century.

Citrus trees were overrun by this new, accidental traveler and unwelcome pest—the cottony-cushion scale. The damage caused by this pest alone threatened to destroy the emerging citrus industry in California. It's important to remember that pesticides were not an available solution in 1889. An effective answer was sought. The eventual solution was drawn from the farmer's knowledge about natural enemies. From close observations in the field, farmers and agricultural scientists became aware that certain predatory beetles, aggressive parasites, and deadly disease organisms could be considered beneficial agents, particularly if they confined their attacks to pests and left the crops intact. This knowledge was shared and soon attracted great interest among agricultural scientists. If we recall history, the writings of Charles Darwin provided some startling new insights into the workings of nature at about the same time. Some of his observations as a naturalist defining the origin of species, raised important questions about the presence or absence of predators in a particular environment. Darwin emphasized the role played by predators as a part of the explanation for the survival or decline of a species. So, in the late 1890s, with the survival of the California citrus industry in jeopardy, and pesticides barely a gleam in the chemist's eye, organic solutions were sought in the workings of nature. It was argued that if nature could accidentally send crop predators across oceans hitchhiking aboard ships or blown haphazardly by the winds, why then couldn't human societies intentionally import their enemies in the same manner. So, when a U.S. Department of Agriculture scout working in Australia found that the vedalia beetle fed on the deadly scale insect, he knew that he'd probably found the natural enemy of the citrus damaging cottony-cushion scale.

At the cost of about $2,000, these vedalia beetles, part of the lady beetle family, were collected by the Department of Agriculture explorer. Scientists then released 129 imported Australian vedalia beetles into the devastated California citrus groves, apparently, just in time. The enemy of the scale began to reproduce and soon spread throughout citrus-growing areas. By the end of the first year of this project, the beetles had dramatically reduced the scale population. Cottony-cushion scales would never again prove to be a serious problem for citrus growers in California.

This technique of releasing an imported organism that establishes itself and spreads permanently to control a pest is known as the classical biological control concept. The economics of this biological control concept are simply remarkable, for not only does the natural enemy of the pest end up saving the crop, but once a predator is released, no further costs are required to keep the pest under control. Later on, of course, scientists realized that

Imported from Australia for its biological control possibilities, the vedalia beetle immediately began eating its way through a troublesome population of citrus scale insects in California. Even before the twentieth century, biological controls were used to save the California citrus crop. (Photo from Agricultural Research Service, USDA.)

many of the beneficial organisms needed to have their populations renewed from time to time. They, too, could be subject to predators or harmful environmental conditions. This augmenting of numbers involved the mass rearing of the desirable or beneficial pests, though costs still remained low compared to pesticides. What is also remarkable is that this biological concept was already in use before the start of the twentieth century and before the heavy application of chemicals became part of the agricultural scene.

By 1912, conservation provided a whole new aspect to the concept of biocontrols. One early conservation project targeted that same infamous boll weevil in cotton. Farmers were advised by agronomists to alternate rows of cotton with rows of a plant that attracted weevils while also containing parasites that would attack the voracious boll weevils.[8] The project might have worked, but for some inexplicable reason, it was never followed through. This was an unfortunate oversight, as weevils nearly destroyed the southern cotton industry. Today, cotton and tobacco farmers in the south are just beginning to employ biological controls after finally realizing that pesticides are ultimately ineffective in solving a good number of their pest problems.

Despite the seeming ambivalence on the part of some farmers early in the century to adopt new methods of pest control, the Department of Agriculture conducted many successful biological control experiments. In 1929, the discovery that the native *Macrocentrus ancylivorus* wasp would attack the newly introduced Oriental fruit moth, brought that problem under control. Ten years later, in 1939, the first commercial microbial pesticide was distributed to attack Japanese beetles. Microbial pesticides are devel-

The boll weevil, which feeds on cotton, flourished in the absence of natural enemies. A U.S. Department of Agriculture program of integrated pest management uses special traps containing weevil sex attractants, called pheromones. The traps are put in cotton fields in late summer. The traps catch weevils and let the scientists know how heavily an area is infested. Next, malathion, a pesticide, is sprayed on fields in the fall, killing weevils before they store up fat and hide under leaves to survive the winter. (Photo from Agricultural Research Service, USDA.)

oped through the use of organisms rather than chemicals. For the past several decades disease organisms have been developed in commercial labs, and from there distributed in the field. However, despite these successes, chemical insecticides entered the scene in the 1940s and, for all practical purposes, stole the thunder from the advances made in biological controls. DDT, in 1939, was the first of many insecticides to dominate the world of pest control.

To the credit of scientists in the Department of Agriculture and in some of the agricultural extension programs, they quietly continued their research on biocontrols and continued to achieve success. At the time, their results went unnoticed, though Rachel Carson made some note of this promising alternative in her book *Silent Spring*.

In 1944, for example, the importation of biocontrol beetles from Australia to control the range weed St. Johnswort resulted in complete control of the weed in California. Biological controls were chosen in these cases, as herbicides and fungicides were still in the developmental stages. In 1957, the importation of parasites against the alfalfa weevil resulted in their control in the northeastern U.S. by 1970. And, in 1964, the introduction of three insects, a beetle, a moth, and a thrips—all of which eat alligatorweed—proved to be a successful innovation. Alligatorweed is a choking weed that grows in water along the banks of lakes and rivers. Its rapid growth eventually clogs the waterways. Since the weed originated in South America, the solution to the alligatorweed problem was sought in its native habitat. About 30 years ago, in the 1960s, Department of Agriculture scientists went to Argentina to find insects that might eat the weed. They came up with the alligatorweed flea beetle, which has practically eliminated the weed in Florida; another insect, a moth, has helped the beetle bring alligatorweed under control in the Mississippi Valley. The third, a thrips, also did its part controlling the weed. The use of biocontrols in this case resulted in control of a major weed in the southeastern U.S. by 1978.

There have been many other success stories involving the application of biological controls, though as pesticides became more commonly used, the tendency was to turn to

Biological controls at work. The South American flea beetle eats alligatorweed, the scourge of many southern waterways. Waterways once clogged now offer easy navigation thanks to the weed hungry flea beetle. (Photo from Agricultural Research Service, USDA.)

killing pests with chemicals. Biological controls tend to be slower and less visible in their effects over time. There is far less support from the government for this type of agricultural research in comparison to the amount the chemical industry allows for pesticide development. Time has been a critical factor in this debate. It takes time to identify natural predators, breed or cultivate them, and release them into the environment. Of course, once they are released, they tend to be self-perpetuating. That is hardly true of pesticides, which require repeated applications. So, in the long term, biocontrols can be both effective and economical.

A seven-spotted ladybird beetle captures a pea aphid, one of its favorite dinners. The pea aphid destroys crops. Aphids, or plant lice, suck a plant's sap, causing it to weaken and die. They attack a wide variety of crops, from alfalfa and apples to roses and wheat. Ladybird beetles are well-known aphid slayers. (Photo from Agricultural Research Service, USDA.)

Recent discoveries are now turning toward disease organisms, such as parasites, to serve as the newest form of predator. But it took 25 years of research (mostly government sponsored) to generate this startling new discovery. J. E. Henry, since retired, and his colleagues at the Rangeland Insect Lab in Bozeman, Montana, have finally placed a grasshopper disease on the commercial market. What this means is that a protozoan, which is a single cell organism of microscopic size, rather than another insect, is being used as a predator. This protozoan, called *Nosema locustae*, kills 50–60 percent of grasshoppers in three to

Tiny (one-eighth-inch long) flea beetle feasting on leafy spurge, a weed of Western rangeland. The flea beetle is one of several biological control agents being tested to combat the costly weed that infests 2.5 million acres of rangeland and pastureland in the Great Plains. (Photo from Agricultural Research Service, USDA.)

Biological controls at work. A Mexican bean beetle larva, one of the most damaging soybean insect pests, becomes a meal for the spined soldier bug. Successful biological control of pests involves attracting and keeping beneficial insects in crop fields. (Photo from Agricultural Research Service, USDA.)

four weeks. It accomplishes this through parasitic depletion of the grasshopper's fat stores which subsequently induces a weakened state in the insect and leaves it without the energy to find food. Thus, grasshoppers die first from the action of the parasite and, second, their death is accelerated by the loss of food intake.

These startling new discoveries can take many years of solid research to unearth. The investment of funds, energy, and resources are what is required to expand the avenues of pesticide alternatives—in other words, there has to be a strong societal commitment to making this

After the heavy application of pesticides over many years, Colorado potato beetles are now resistant to most registered insecticides and can cause severe damage in some crops, notably potatoes. Through research in biological controls, the U.S. Department of Agriculture has experimented with insect-resistant genetics. In this photo, Colorado potato beetles take only one or two bites of this insect-resistant potato before they are repelled. The potato has been genetically engineered to contain a rare gene found in wild potatoes. (Photo from Agricultural Research Service, USDA.)

necessary transition in our agricultural practices. Unfortunately, in the twentieth-century, biological controls have always been used as a second choice. The tendency has been to choose pesticides as the first course of action. The case of the Mediterranean fruit fly is an excellent example of the social complexities influencing the dilemma of pest control practices in the latter part of the twentieth century.

THE DILEMMA OF THE MEDITERRANEAN
FRUIT FLY

The Mediterranean fruit fly is an ongoing twentieth-century problem. Much like the desert locust who crossed continents on the winds of a hurricane, the "Medfly" is another excellent example of a pernicious hitchhiker. Of course, the more travel going on between countries and continents, the greater the likelihood of transporting undesirable insects and plants. In our fast-moving society, travel has accelerated through air transportation. Thus, hitchhiking insects have greater opportunities to travel into regions otherwise unknown to them. The Mediterranean fruit fly, whose name suggests its origins, has been plaguing the California fruit crops on and off since about the 1960s. In the winter of 1990, for the second time in this decade, California has undertaken vast aerial spraying of the pesticide malathion over residential areas to eradicate the fruit fly. There has been a public outcry over the widespread use of malathion by nearby residents of the area, whose backyards would be covered by the pesticide left by aerial spraying. Apparently, they are not convinced of the chemical's safety, though some agricultural scientists suggest that malathion is a least-harmful pesticide. In fact, malathion is responsible for honey bee kills where it is sprayed, though the fear of the chemical comes from what is unknown about it rather than what is known. The honey bee kills do demonstrate that some pesticides can destroy harmless and innocent bystanders, in this case the bee, an important player in the pollination of plants. So, malathion can be considered less harmful in comparison with other pesticides, but that in no way suggests that it is entirely safe. But is there an alternative? After all, Medfly infestations of California's recent past have been known to destroy $100 million worth of crops in a single year. In fact, there is a solution drawn from a system of biological controls. Why then was it not used? Biological controls had been in place in sufficient time for the possibility of another outbreak of the Medfly. Unfortunately, these non-chemical controls were insufficient in numbers to employ at the time of the insects' sudden appearance. The strategy is well tested, though. By breeding and releasing sterilized male fruit flies, the Department of Agriculture can naturally diminish an insect population. Releasing the nonreproductive males into the fields allows them to mate without producing offspring. Eventually, the insect is severely curtailed. In this particular Medfly event, while sterilized male fruit flies had been cultivated in California, they proved insufficient in number and time to complete the cycle for sterilized breeding. Hence, with no other known system of deterring the damage caused by the flies, the state government, led by Governor George Deukmejian, concluded that repeated aerial sprayings of malathion were in order. Those who were against the strategy were the local homeowners. In a particularly harsh appraisal of those local malathion opponents, a *New York Times* editorial stated that "California's suburbanites are being too selfish in trying to avoid the harmless inconvenience of a malathion shower in their backyards. They, too, stand to gain from the saving of the state's harvest."[9] Malathion is referred to in this same editorial as a "mildly toxic sticky goo." The *Times* then points out that

> the public already accepts tiny residual doses of pesticide in its daily diet as the price of a cheaper, maggot-free food supply. It's not very convincing when California's subur-

banites protest malathion spraying that probably adds little or nothing to this everyday exposure.[10]

Strong evidence in favor of saving the crop by the use of
multiple sprayings of malathion seemed, at least to the
Times, more compelling an argument than the concerns
expressed by suburbanites. Certainly this debate calls for
a risk assessment. The only problem is that the data available on malathion health effects is too skimpy in comparison with the hard facts known of past economic disasters
resulting from Medfly-induced crop failures. So while we
do not know what possible harm can come from exposure
to malathion in this case, we do know in very hard dollar
terms what losses can come from the unchecked presence
of medflies in the California orchards. This inbalance in
the weighing of evidence leaves the community in question boxing shadows or ghosts when dealing with their
health concerns. Their negative reaction may also suggest
that people are starting to become aware of the potential
dangers of pesticide exposure. At this point, if malathion
is not a problem, they cannot know that to be the case any
more than if it is.

What makes this one tiny fly so unwelcome to U.S.
shores? The Mediterranean fruit fly is a crop-destroying
fly that lays its eggs in 250 kinds of fruits and vegetables,
leaving them unfit for sale. It is a persistent and frustrat-
ing problem. And there are others of this type to ravage
the abundant California crops. An infestation of Oriental
fruit flies was suppressed in 1974, and a later Mexican fruit
fly invasion in 1984. The problem is seldom ever eradicated, even with the spraying of pesticides, for they, the
crop damaging insects, come alone—unencumbered by
their natural predators, who probably kept them in check

in their original habitat. Free to devour crops in the new land, these immigrant pests can suddenly and disastrously change the course of agricultural history. It is important to note that there does exist an alternative to extensive pesticide use through the application of biocontrols; yet in this most recent outbreak of Medflies in California, it was not effectively available when the moment of crisis arose. From this we understand that the use of biocontrols requires considerably more advanced planning than the relatively quick application of pesticides.

Two things become clear when we examine the changes biocontrols made. First, there is a period of time necessary to effect natural controls. From the introduction of the insects to control alligatorweeds in 1964 to the effective control by 1978, we see a time lag of 14 years. While that may seem a long time to a farmer whose crops are on a year-by-year profit scale, we must also remember that the weed control ultimately affected an entire region of the United States. That is a large portion of agricultural land to consider changing. So many of these benefits from the application of biocontrol technology are available today, provided state and federal agencies engage in advanced planning to put the process to work. This requirement for advanced planning is the critical difference between using biocontrols or using pesticides. Second, we can see that biocontrols, once successful, become a self-perpetuating solution. The economic benefits to the farmer are substantial compared to the repeated need to spray pesticides, which in the long run may not prove effective. Finally, biological controls of agricultural pests is largely the effort of one or more government agencies. As such, both the government and the public tend to carry the costs of research and education of farming communities. Funding for government programs come ultimately from public taxes, so in this sense, the public is asked to carry some of the burden of changing agriculture—though its not really a burden if the food growth is environmentally safe and healthier for people. However, farm subsidies could be channeled into supporting biocontrol programs instead of their present agenda. Farm subsidies attempt to create a balance in the marketplace. That funding is substantial and could be partially redirected toward the research and application of healthful farming technologies.

THE COST–BENEFIT RATIOS OF APPLYING BIOCONTROLS IN LIEU OF PESTICIDES

The cost–benefit ratios for classical biological control are known to be highly favorable but are often impossible to compute accurately. However, a few published examples of estimated savings resulting from complete or substantial biological control do exist.[11]

1. Annual savings of $4,000,000 for control of the black scale and citrophilous mealybug on citrus in California.
2. $8,000,000 a year savings for control of the alfalfa weevil in the middle Atlantic states. This is both an early and a very low estimate. The annual savings from 1985 to 1986 was estimated to be $48,900,000.
3. $17,000,000 a year reduced turf maintenance costs due to control of the Rhodesgrass mealybug in Texas, with additional benefits through increased calf sales of $177,000,000 up to 1978.

4. $13,000,000 savings per year for control of the alfalfa blotch leafminer in the northeastern U.S.

5. $22,000,000 savings per year over previous control costs plus weight gain for cattle, due to control of common St. Johnswort in California. This was accomplished through the application of bio-controls. The Klamathweed beetle, imported from Australia, was turned loose to eat up a troublesome weed in the western U.S. This weed, called the Klamathweed, St. Johnswort, or goatweed, is poisonous. It was spreading out into California rangeland where cattle would accidentally eat the weed. In the mid 1940s, when the beetle was first turned loose on U.S. soil, this weed was aggressively spreading across rangeland from northern California to southern British Columbia in Canada. Within 10 years, the now thriving beetle population had reduced the yellow-flowered weed by 99 percent. The estimated savings to ranchers was approximately 22 million per year. From combined efforts of scientists from the U.S. Department of Agriculture and the University of California, the importation of the Klamathweed beetle was apparently one of the first attempts in the continental United States to control a weed with a plant-feeding insect.

6. At least $400,000 annual savings in treatment costs due to control of the alligatorweed in the southeastern U.S.

In a book aimed at agricultural scientists and written by scientists from the Agricultural Research Service, Jack R. Coulson, from the Beneficial Insects Laboratory in

Maryland, and Richard S. Soper, leader for the Biological Control Room, make the following observations: "University of California scientists estimate a $30 return on every $1 expended on their biological control importation programs and a total savings to the agricultural industry of California of $274,942,000 that accrued from 1928 to 1973."[12] Coulson and Soper conclude, "these cost/benefit figures for classical biological control are considered to be very conservative estimates, but are obviously very favorable, especially considering that these savings accrue annually."[13]

Most important to this discussion is the fact that each pest controlled by its natural enemies results in less use of chemical pesticides in the environment. Since the diminishing of chemical uses in the environment is a major goal of transforming agriculture, the economic benefits can only add to its attractiveness. Still another facet of biological controls is the use of *pheromones*, which include insect sex attractants. The scientist can analyze the patterns of sexual activity in specific insects and disrupt their mating patterns by filling the air around an infested tree with female insects' pheromones. In a five-year study of several peach orchards in central Georgia, scientists with the Agricultural Research Service at the Gainesville, Florida, Insect Ecology Research Laboratory achieved almost complete control of the peachtree-destroying peachtree borer by hanging a pheromone-containing dispenser in each tree. As the male futilely attempts to mate with the non-existent female, he becomes confused and exhausted. These synthesized pheromones can prevent mating and reduce populations of pests to practically zero. The economic benefits of using pheromones are considerable when we consider that these borers cost Georgia peach growers some $1.6 million a year in crop losses and control costs. Nationwide, commented the Agricultural Research Service entomologist, J. Wendell Snow, the problem costs about $20 million annually.[14]

From an economic standpoint, the biological control activities stay within the domain of government agencies and agricultural research stations. As such, private industry does not have a significant role in the process (nor an industry as such from which to gain profits). Economically, nothing can match the enormity of wealth generated by the chemical industry. What it does not sell domestically is certainly exported to other, less pesticide-regulated countries. In conclusion, we can see that biological controls save on costs in a number of ways, but they do not generate capital in the same manner that private industry does. Biocontrols, after all, fall almost completely in the province of government.

BIOCONTROLS AND GOVERNMENT: A MATCHED PAIR

Two U.S. Department of Agriculture agencies work together to find, collect, test, and disperse biological control agents: The Animal and Plant Health Inspection Service (APHIS) and the Agricultural Research Service (ARS). A biological control advisory group selects projects with the most promise of success, and federal specialists raise the needed quantities of beneficial organisms. State governments and industry groups help disperse the organisms, following the federal plan of operation. This approach to pest control is clearly not a private industry. In that sense, the profit motive is absent; but the profit motive is certainly present when the savings affect the agricultural industry. A good example of the economic benefits to farmers is the case of the cereal leaf beetle. With the introduction of tiny parasitic wasps as natural enemies of the beetle, $14 million a year was saved. The total project cost $14 million, but saved the farmer $14 million per year from losses due to damages from the beetle.

The alfalfa weevil was brought under control by the use of beneficial parasitic wasps raised by a research station in Niles, Michigan, and shipped to other parts of the country. What is significant about this case is that the beneficial wasps would normally have been destroyed by powerful pesticides used to control the weevil.

But not all problems can be solved by the importation of natural enemies. The grasshopper is one of those cases.

From his U.S. Department of Agriculture research center in Boise, Idaho, Jerry L. Fowler has been working with the U.S. Department of the Interior, the U.S. Environmental Protection Agency and six universities to "place grasshopper management on the cutting edge of technology." Though Fowler acknowledges that grasshopper infestations may never become eradicated, the technology being developed by the Grasshopper Pest Management Project will enhance outbreak prediction ability, management decision-making, environmental preservation, and provide new management options. But pesticides have not been omitted from the list of management options. Malathion, noted before as the Medfly insecticide in the latest California controversy, is considered a less harmful insecticide. This chemical is recommended for use, though as mentioned before in the Medfly case discussion, it has been known to cause bee kills. A search for alternatives to pesticide use is turning us further toward research in biocontrols.

Despite the continued use of pesticides in the control of the nearly unpredictable grasshopper infestations, progress is happening in the realm of pest management, if not control. Management techniques make use of climatic and biological observation sites as a means of anticipating the onset of an infestation. Since it is known that certain climatic conditions, such as heavy spring rains, can help to trigger the appearance of grasshoppers, a monitoring of the climate and biology of an area is an effective way of anticipating a possible infestation. But efforts also continue in the Department of Agriculture to develop biological control agents for the persistent grasshopper.

INTEGRATED PEST MANAGEMENT

Biological control of pests is actually one part of a larger concept known as integrated pest management (IPM). The relatively new strategy is based on determining an economic threshold that indicates when a pest population is approaching the level at which control measures are necessary to prevent a decline in net returns. In other words, a search for alternatives when pesticide use is no longer economically feasible as a single control mechanism. In principle, IPM is an ecologically based strategy that relies on natural factors, such as natural enemies, weather, and crop management, and seeks control tactics that will disrupt these factors as little as possible. For example, farmers can reduce pesticide use on cash grains through crop rotations that disrupt the reproductive cycle, habitat, and food supply of many crop insect pests and diseases. If this all sounds familiar, it is because this important agricultural knowledge was available and to some extent in use before the advent of pesticides. But because today several generations of farmers have relied on the regular application of pesticides (whether needed or not), this knowledge and skill have been both ignored and forgotten. Alternative farming practices therefore require more information, trained labor (learning to recognize the appearance and signs of a problematic pest, for example, which would offer more time to organize control strategies), and management skills per unit of productions than what is needed for conventional pesticide farming. In a sense, conventional farming has come to mean the regular use of pesticides on a seasonal basis without investigating the presence of crop enemies or the extent of the need for chemicals. Oddly enough, many federal government policies discourage adoption of alternative practices by economically penalizing farmers who adopt rotations or attempt to reduce pesticide use.

This occurs because support is given by the federal government for the growth of certain designated commodities. But, if production of those crops are interrupted by the alternating system of rotation of crops, or diminished in volume by a loss of crops by suddenly withdrawing pesticides, then the farmer does not receive the financial support. It is paradoxical that the same government that supports research in biological controls also encourages unrealistically high-yield goals, inefficient fertilizer and pesticide use, and an unsustainable use of land and water. The federal programs we are referring to are the commodity price and income support programs.

GOVERNMENT POLICIES THAT DISCOURAGE AGRICULTURAL CHANGE

Federal farm programs are designed chiefly to support crop prices and farm income. It may well be the tail that wags the dog in the planning of agricultural methods. The program supports the growth of specific crops. Then, commodity price supports are established by the government, direct payments and crop loans are made to farmers, and agricultural land is diverted from production through paid land diversions. In effect, the specialized cultivation of one or two crops is rewarded over the varied rotation of crops. Yet, the use of rotations would discourage the population growth of pests that feed on specific crops; and if a rotation involved the planting of a legume known to supply nitrogen and conserve the soil, it would be an environmentally sound method as well. If, however, that same legume were not on the list of specified and supported crops, it would be less financially rewarding to the farmer in the short term. In the long term, our environmentally unsound practices would add up in other costs from soil and water depletion. So, the practice which encourages crop specialization over large acreages also increases the likelihood of needing and using pesticides as well as employing environmentally damaging fertilizers. If farmers reduce the acreage devoted to the highly rewarded commodity program crop, such as corn, feed grains, wheat, cotton, and rice, for two or three years in order to rotate to another crop, they also reduce the acreage eligible for federal subsidies. And as in other matters of commerce, money talks.

A second point of government regulation which runs

counter to employing IPM techniques is the problem of the federally approved grading standards for fruits and vegetables. These grading standards dictate that the most cosmetically attractive crops can be marketed as raw commodities. The more damaged and blemished products are directed toward processing and canned goods. But the price return is higher for the raw fruit and vegetable market. These grading standards usually allow few surface blemishes on fresh produce or extremely low levels of insect parts in processed foods. As already noted, pesticides can offer great cosmetic value to fruits and vegetables. Alar, for example, helped apples to remain on trees until they ripened into a deep color. That is what the public has come to expect. So, in order to receive a top price for produce, farmers have to use more and more chemicals to meet the new standards.

All in all, these federal incentives, which encourage high yields of specified crops per acre, also unintentionally discourage farmers from employing land and water conservation. Yet, despite these financial short-term considerations, some farmers have begun to apply IPM techniques. They are few in number, and most of these farmers run smaller-scale operations. They are not the massive agribusiness corporations. In the Midwest to some extent, and in upstate New York, there is the continued presence of the independent family farms operating with less of a profit margin. A few of these farms are employing crop rotation, conservation, and the use of biological controls. Still, IPM is not the same concept as organic farming. The latter idea employs no pesticides whatsoever (if it is truly organic). This organic method employs only biological controls, while IPM may use chemicals as considered necessary, but not to excess, and with specific targeting of pests. Consumers have begun to notice an increase in the availability of organic produce in small markets, and the interest in purchasing food without the presence of pesticides is on the rise. And more encouragement for alternatives in agriculture comes from the continued scientific findings illustrated by the success of biological controls. In some recent cases, biological control performed a service to farmers that pesticides simply could not accomplish; this in itself should contribute to the growing enthusiasm for changing agriculture. The case in point is the National Heliothis Suppression Program.

CONTROLLING THE CORN EARWORM AND THE TOBACCO BUDWORM

The corn earworm (*Heliothis zea*; also known as the cotton bollworm or tomato fruitworm) and its sister species, the tobacco budworm (*Heliothis virescens*) were both devouring around $1 billion in cotton, corn, soybeans, tobacco, tomatoes, and many other vegetables each year. Why didn't they bring on the pesticides, you might ask? In fact, all of this devouring was going on despite the fact that pesticides were being applied to these crops annually at the cost of $250 million. That fact alone explains the incentive to adopt probably one of the most complex pest research programs to date.

In 1986, the Agricultural Research Service (ARS) of the U.S. Department of Agriculture consolidated its efforts on these two pests and made it an ARS National Program. Six major control approaches entered into the research:

1. *Genetic control via sexual sterility.* By crossing the tobacco budworm with a related species, scientists came up with a strain that consistently produced all sterile males. Scientists also found a way to substerilize bollworms with a dose of radiation so that they only produce male offspring.

2. *Biological control.* Diseases, predators, and parasites can do their part in controlling these pests. Currently, scientists are genetically trying to manipulate corn to incorporate the insect toxin of a bacterium. In this approach, the plant is naturally harmful to insects that would normally feed on and destroy them with impunity.

3. *Behavioral control.* To prevent mate-searching moths from finding each other, scientists have isolated the chemicals that inhibit enzymes vital to pheromone production. Pheromones are the sex attractants that insects give off to accomplish mating. No enzymes means no pheromones, and no pheromones means the moths will have no way of finding each other to mate and reproduce.

4. *Chemical control.* This research has sought to find new, environmentally compatible chemical insecticides for industry. In addition, scientists have sought new ways to release chemicals slowly into the fields, which would kill the insects more selectively and efficiently. The slower the release of chemicals, the more they can affect large numbers of insects over a period of time without needing to spray repeatedly. It is the repeated spraying and exposure to chemicals which presents a health danger. The current use of pesti-

The parasitic wasp lays her eggs in a tobacco budworm (*Heliothis virescens*). By putting this natural predator to work, scientists hope to control members of the genus *Heliothis*, which cause a billion dollars in annual damage to cotton, corn, soybeans, tobacco, and other crops. (Photo from Agricultural Research Service, USDA.)

cides involves widespread and repeated sprayings to be effective. In another approach, putting insecticide in oil- instead of water-based materials keeps them from evaporating quickly. The benefit is that the chemical remains effective over a longer period of time. A possible consideration is the effect upon farm workers. As mentioned in Chapter 3, when properties were added to a pesticide formula in order to enhance the period of effectiveness, the reentry period needed for workers to safely return to the fields has to be taken into consideration.

5. *Host-plant resistance.* This strategy involves the release of new crop lines, including corn and cotton, with various levels of resistance to pests.

6. *Cultural control.* This involves the plowing up of stalks or otherwise getting rid of crop residue to kill dormant and developing insects in the soil. Scientists are studying as well what roadside plants these pests like to feed on to try to destroy food sources they use between crop plantings.

All of the above research is part of the repertoire of techniques used toward achieving the goals of IPM. The concept does not eliminate the use of chemicals altogether, but certainly they constitute only one in six possible routes taken in pest control. In many cases multiple approaches could be employed concurrently. Scientists have been ingenious in developing new techniques for applying technologies that would be most effective on a particular crop. For example, they use pheromone traps and echolocation devices to count insects. (One of the great unknowns in pest control is how many of them are out there anyway. If we knew that, we could determine not only what should be used as a control, but how much is necessary.) So research scientists are finding out exactly what pest level warrants pesticide use. They are learning as well to count beneficial organisms in a field to see if there are enough to provide pest control without insecticides or with less insecticides.[15]

An Iowa farmer, Dick Thompson, makes use of ridge-tillage by planting in rows of unplowed, raised beds. This type of land tillage elimi-
nates the need for fall tillage and spring preplant tillage. Because he
uses no plowing to stir up the seeds of weeds, the farmer controls the
growth of weeds without using herbicides. (Photo from Agricultural
Research Service, USDA.)

Does all this scientific activity cover the needs of
changing our agricultural practices? Indeed not. There is
one further avenue of change still necessary. Pests and
weeds do not carry passports, nor do they speak a separate language when they cross from one state to the other,
or from one border of North America to the other. Insects
can hop, fly, or hitchhike across seas and continents. The
cotton boll weevil, for example, "came a visiting" from
Mexico in the early 1900s.

Controlling the pest problem when they cross borders
from one country to the other involves the cooperation of
countries to the north and to the south, and this is the goal
of the North American Plant Protection Organization.

CANADA, UNITED STATES, MEXICO

The North American Plant Protection Organization, NAPPO, serves Canada, the United States, and Mexico by sharing information and establishing goals in plant health activities. NAPPO was created in 1976, that active era of environmental concern which brought so many agencies and legislative actions into being. The organization comprises federal plant protection officials from each of the three countries and affiliates with the United Nations' Food and Agriculture Organization, FAO.

While the initial interest of the NAPPO partnership was to control and prevent the spread of plant pests, such as the boll weevil, across international borders, their more recent projects have added the exchange of technical information on effective survey, regulatory, and pest management procedures. They also act in the interest of plant protection by developing uniform minimum standards in North America for quarantine activities such as inspection of commodities at ports of entry, which is precisely when hitchhiking insects and weeds arrive, and inspection for export, when undesirables leave to plague other nations.

At a recent conference of the Entomological Society of America, the Mexican speaker, Eleazer Jiminez, noted that "most world countries are working intensively—and for some time now—into the possibility of utilizing more broadly natural enemies of plant pests from any biological origin (parasites, preditors, virus, bacteria, fungi) . . ."[16]

The Canadian speakers, C. D. J. Miller and J. S. Kelleher, both from the research branch of Agriculture Canada in Ottawa, were proud to point out that biological control in Canada had started in 1882. In the early twentieth century, Canadian scientists investigated the larch sawfly, the browntail moth, the oystershell scale and the wooly apple aphid which were acclaimed as biocontrol successes and led to the appointment of full-time biological control workers and the establishment of a Parasite Laboratory at Belleville, Ontario, as early as 1929. Early progress in biological controls may well have been overshadowed by the advent of powerful pesticides, but recent protection programs in the research branch of Agriculture Canada have biological control research components associated with them.[17]

CHANGING AGRICULTURE TODAY

Changing our agricultural practices is the key to changing our chemically dependent society. We can see that the changing process itself is complex and a difficult path to follow, but certainly not impossible. There is government-supported research and implementation of nonchemical biological controls, but there are also government price supports for farmers that inadvertently lead to a greater and greater use of pesticides. If the two seem contradictory—it's because they are. They would be less so if price supports would also be used to encourage and benefit the research and implementation of biological controls in particular, and of IPM technology in general. Society is always changing, and we often find ourselves living with the decisions made by earlier generations and administrations. What we need to understand at this point is how society and its agricultural practices can arrive at a more healthful resolution to the demand for clean food and water.

CHAPTER SEVEN

THE GROUNDSWELL

Demanding Clean Food and Water

In 1970, the world's first "Earth Day" proved to be the largest organized demonstration in all of America's turbulent history. More than 20 million people took to the streets, united in their cause: to save our polluted and threatened environment. With nature walks and "teach-ins" at college campuses, students and their equally concerned professors held informal seminars and "rapped" about saving our globe from what was rapidly becoming an unlivable environment. Campuses across the country and local streets became fairgrounds, with booths promoting environmental concerns. Young men and women handed out a single blossom, a daisy, a flower which had only recently been the symbol of an electrifying student peace movement against the Vietnam and Cambodian wars, but had now become a symbol for preserving the earth. The federal administration, wary of student movements in general, was rumored to have sent the F.B.I. to photograph participants of Earth Day to record their "revolutionary" activities. But these students were not alone. In deference to the celebration, the mayor of New York City banned automobiles from Fifth Avenue (unheard of at the time), and 100,000 people attended an eco-fair in Union Square. To demonstrate their support of Earth Day, the U.S. Congress formally adjourned to allow members to attend teach-ins in their districts.

That startling decade of the environment, which led to the formation of the Environmental Protection Agency in 1971, and a series of legislative acts, starting with the Clean Water Act, was the beginning of a groundswell of public activism in behalf of the environment. That activism lasted nearly the entire decade until the 1980s brought us into the world of the "me" generation and the self-involved, socially disengaged Yuppie (although the Yuppie alone cannot be held responsible for our national amnesia concerning the environment). It is not surprising that during the 1980s the state of our environment again declined at an alarming rate. The concerned decade of the 1970s had taught us that politicians had the power to change our laws, and that people could push politicians to make those necessary changes. This participation of citizens was integral to the survival of the partnership. As we entered the 1980s that partnership dissolved as we lost our social and environmental consciousness—citizens and politicians alike.

The concept of Earth Day as a celebration of environmental consciousness fell into a dormant state—a hibernation of a decade-long winter. It did not reappear until the new era of the 1990s. Though it entered with the same fervor and enthusiasm that ushered in the 1970s, there were certain, noteworthy differences, given the time gap of 20 years. The most important change was that the concern for the environment was now dominant within mainstream America. This groundswell of support that pervaded all regions and sectors of American society to improve the environment was totally unprecedented. No longer were environmental issues associated with the radical social movements that characterized the politics of confrontation in the 1970s Civil Rights movement, the student movement, the womens' movement, and lastly, the environmental movement. Earth Day 1990 was an omnibus that nearly every segment of society wanted to board.

Flanked by United States Senators, Members of Congress, and representatives of almost every major American environmental group, Denis Hayes, the chairman of the Earth Day 1990 coalition, called for the "worldwide mobilization of a massive citizen army to avert planetary disaster." Providing a line of continuity in history, both Hayes, who was coordinator of the first Earth Day in 1970, and Gaylord Nelson, the honorary cochairman of Earth Day 1990 and father of the first Earth Day, provided the visible leadership. For several months before the April event, Nelson spoke publicly of the unfinished job started in the 1970s: "Clearly our job is not done. The threads of the net that hold the world ecosystem in balance are breaking and unraveling. Only a huge coordinated, worldwide effort will save what is left of the natural world."[1] Hayes was more dramatic in his message: "The earth is on the verge of a breakdown. The time to act is now. Millions of 'footsoldiers' are needed if we are to save the planet."[2] But beyond the footsoldiers based on college campuses across the United States were the mainstream leaders of industry, labor unions, media, and political luminaries and the ubiquitous leaders of cities and states. Some had been involved with environmental issues for many years;

others were only recent joiners. The Earth Day 1990 board of directors, a diverse, well-rounded group, included the Texas Agriculture Commissioner Jim Hightower, Ted Turner of Turner Broadcasting, minority leaders Eleanor Holmes Norton, Jesse Jackson, and Denver mayor Federico Peña.

The breadth of Earth Day 1990 spanned critical global issues, more far-reaching than those emphasized 20 years before. There was a clear recognition that environmental as well as energy problems were international in scope. Citizens focused their goals on banning chlorofluorocarbons that destroy the ozone layer, halting the export of toxic wastes and dangerous pesticides to the Third World, beginning a 20-year transition from fossil fuels to renewable energy sources that do not contribute to global warming, reducing acid rain by 80 percent, and creating a strong international agency with the authority to protect the atmosphere and the oceans.

Countries from as far away as Africa, the Middle East, Asia, the Pacific, Central America and the Caribbean, North America, South America, and Europe had been contacted and responded with enthusiasm and involvement—some more than others. The North American continent was strongly involved. Canada brought together an influential board of directors composed of environmentalists, writers, and business people. Media experts, strongly represented in every board of the newly spawned environmental organizations in the United States, were developing television and newspaper campaigns. Oddly enough, some of the long-standing writers and researchers who had worked for decades on environmental issues were not as much in evidence. They were eclipsed by the personae of movie stars, familiar faces, smiling out at us from the television screen. As much as anything, in keeping with changing times, Earth Day 1990 promised to be a spectacular media event. But that is not unusual for our era—television alone has altered our political process in electing candidates and understanding social and political concerns. Sophisticated telecommunications technology through computers and the fax has opened up international networking systems. These technological changes hasten the way major citizen events are organized.

THE INTERNATIONAL NETWORK

As the word for April 22, 1990, Earth Day spread, each country adopted the event to meet its own particular concerns about their environment. China, for example, with its extremely large population problems had great interest generated by their family-planning associations. In Hong Kong, religious philosophers of the ancient Buddhist Perception of Nature made plans to join up with the country's modern, governmental Department of Environmental Protection. Added to the group was the environmental organization Friends of the Earth and the Body Shop (a new California business that specializes in marketing natural products for cosmetics and drugs). In Argentina, Amigos de la Tierra (Friends of the Earth) planned to link Earth Day events with their own national concerns over nuclear waste dumping.

Most of the activities planned for Earth Day appeared to be more symbolic and global than substantive—but that was true for the original event as well. Tree planting was then and is now a ritual included in nearly every celebration of the earth. Earth day 1990 was no exception. The cry in 1970 was simply a general call to clean up the water and the air; nothing more specific than that was declared on that first of all Earth Days. But that symbolic cry created the pressure for the drafting of actual laws that followed.

In this decade's worldwide Earth Day celebration, clean-up campaigns, poster contests, lectures and seminars, concerts and performances, and collection drives constituted the environment awareness activities. The Netherlands organized its Earth Day event on the subject of organic gardening and ecological food, but they were singular in narrowing the event to this one specific issue.

Regionally, in the United States, both the small rural and urban areas offered their own a unique approach to improving their environment. While Newark New Jersey's Mayor, Sharpe James, planned to plant some 2,000 new trees to line the streets in the city in 1990, he also offered substantive changes in legislation that would affect the Newark environment for some time to come. Newark has adopted, implemented, and enforced its own ordinance banning the sale and/or use of polyvinyl chloride and polystyrene food packaging in all of Newark's approximately 1,300 retail food establishments. The ordinance also requires that all retail food packaging must be degradable, though an exemption is provided to retail food establishments which recycle a minimum of 60 percent of their food packaging. The mayor also implemented a citywide recycling program, with an effective recycling rate of over 41 percent of its municipal solid waste.

In rural America, a national survey of rural and small town Americans on environmental issues revealed that these citizens are pessimistic about the nation's environment. They understand that hazards like polluted air, dirty water, and toxic waste are no longer just urban problems, according to Senator Patrick Leahy of Vermont, and Linda DiVall, president of American Viewpoint, a leading Republican pollster. One of the key issues explored in this national survey of rural and small town communities is the question of regulating pesticide use. The survey revealed that rural Americans strongly support an increased government role in regulating pesticide use, despite the potential economic impact that this might have on farmers and the agricultural industry.

How was this environmental consciousness suddenly reborn? For the past 20 years, only the diehard environmentalists and scientists have continued to research the host of environmental problems plaguing our nation. Issues concerning the environment and energy resource management were merely ghosts of the past to the majority of the population. These problems had not gone away—to the contrary, they had worsened during our years of indifference. The answer to the question whence the groundswell of public concern for the environment lies in the series of events which have surfaced over the past two or three years and reported by the media. The public has found these reports to be very alarming. They include: tainted shores, beach closings, and medical waste washed ashore; the carnage of dead and dying dolphins spread along the eastern coast of the United States from the Carolinas to New Jersey; the unwelcome news that the eroding ozone layer poses greater risk of skin cancer; that the greenhouse effect is no longer just a speculative idea, and that acid rain is destroying forests, lakes and rivers from the United States to Canada, and all across Europe, wherever industrial activity and auto emissions could add to the pool of chlorofluorocarbons. And then came the oil spills destined to despoil the wildlife and waters of Alaska, Delaware, New Jersey, New York, and Rhode Island. The

Alaska oil spill was perhaps one of the problems that irked Americans more than the omnipresent toxic waste sites. These are more or less global problems, and their dimensions are still not fully known. We do know that America has a serious toxic waste problem, and its dimensions are still largely unknown.

Kurt Fensterbush and Jon Kerner presented a paper in 1988 on the "Examination of Citizen Associations Against Contamination As Local Movements for Political Change" at the Southern Sociology Meetings, and they point out that in 1985, the Office of Technology Assessment estimated that there may be 10,000 hazardous waste sites in the United States that threaten public health and should be cleaned up or more adequately contained. The costs, the authors contend, could exceed $100 billion. Moreover, the General Accounting Office estimates that over 378,000 waste sites may require corrective action. To date, little corrective action has occurred.[3]

THE SUPERFUND—CLEANING UP THE ENVIRONMENT

The federal program for dealing with the most dangerous of these waste sites, commonly referred to as the "Superfund," is The Comprehensive Environmental Response, Compensation and Liability Act, (CERCLA), a piece of legislation passed in 1980, after more than 10 years of alarming, environmentally based health hazards had surfaced. It was a troubled act from the start: Responsibility was delegated to the industrial private sector to pay the bill for clean-up operations. Problems of enforcement were compounded by the emergence of this unusual piece of legislation during the Reagan era—a time when the environment was not a high priority. Therefore, after its first five years, the Superfund had only cleaned up six sites, and those were considered extremely easy to clean, or else the clean-up operation had not fully stopped the problems. This outcome was extremely disappointing to the Superfund-targeted communities where people had built their homes and lives, and where they felt a right to expect a clean and safe environment. Over a period of 30 or even 40 years, the byproducts of industrial wastes and agricultural pesticides had ended up in local streams, lakes, and rivers. This surface and ground water contamination led to the destruction of many local water supply systems.

The Superfund was first looked upon as the key to solving these environmental problems, but the frustration of local citizens in their efforts to obtain even the most cursory information about the possible health hazards in their contaminated water supply led those community people to conclude that the federal government, and most particularly the Superfund, was a mere tangle of slow-moving red tape.[4] The frustration of communities to gain even the most elementary information about the chemicals they were imbibing, the danger to themselves and their families, and the health problems that were simply not documented, despite growing evidence, did finally force them to organize into grassroots, community-based groups.

The names they chose to call themselves as community groups clearly stated their own mission and concern. With acronyms such as PURE (People United to Restore the Environment), FACE (For a Cleaner Environment, Inc.), CAPE (Citizens Against a Polluted Environment), and simply HELP (Helping Eliminate Land Pollution), the groups were largely formed by people who had never before joined an activist organization, and who had never marched in any public demonstration other than, perhaps, the Thanksgiving Day Parade in their own home town. While these regional groups were mushrooming across the country, others were forming nationwide movements to deal effectively with single-issue problems. One of those problems was the question of pesticides' presence in food. While many communities took on the issue of hazardous waste polluting their water, other groups, particularly mothers, banded together to protect their children from fruits laced with pesticides. The effects of this groundswell of concern was formidable.

MOTHERS AND OTHERS AGAINST PESTICIDES: ALAR AND APPLES

"Score one for consumers in the quest for a safer food supply," begins an article in the premier, Volume 1 edition of TLC, an acronym for "truly loving care," published by Mothers and Others for Pesticide Limits. In the newly formed newsletter, the consumer group, a project of the Natural Resources Defense Council, continues their campaign against Alar, a growth regulator of apples. The article continues:

> Because of strong public pressure, the apple industry announced in May [1989] that U.S. growers would stop using the chemical growth regulator Alar on their crops. Soon afterwards, Uniroyal Chemical Co., which manufactures Alar, decided to take the chemical off the domestic market altogether.[5]

This is good news for parents whose children consume large amounts of apple products like juice and applesauce, the article explains, for "its been known for years that the chemical UDMH, formed when Alar-sprayed apples are processed, is a potent carcinogen." Indeed, public outcry about the potential health effects of Alar and its breakdown chemical was probably the single most important factor in the withdrawal of Alar in apple growing.

By the spring of 1989, full-page stories about the potential health effects of Alar had reached most major U.S. newspapers. The apple industry, according to the *New York Times*, citing a sharp drop in sales that had already cost growers $50 million, would voluntarily stop using the chemical Alar by the forthcoming fall.

> At a news conference here, spokesmen for the apple industry said that consumer reaction to news reports about the risks of Alar had cost the industry $50 million so far this year and that growers and processors would probably lose another $50 million by years end.[6]

Economics of this proportion was clearly a deciding factor in the swift withdrawal of Alar, first from domestic crops, and later from exportation. In early 1990, Uniroyal went beyond halting domestic sales: it confirmed the withdrawal of the chemical from overseas sales to fruit- and vegetable-growing countries.

How did the EPA respond to the Alar issue? It appeared to be waffling, first stating in the winter months that only 5 percent of the crop was treated with Alar and, later, in May, revising its estimate to 15 percent of the crop. What did this mean? Had earlier estimates been quoted to underestimate the extent of the problem? We can only guess about why and how these figures were revised.

ALAR AND DDT: A WORLD APART

The withdrawal of Alar from apple growing happened in a few short months. This is a dramatic case when we consider that DDT, a more persistent and ubitquitous chemical, took three long years of intense battle before it was banned from use. But there are important differences in the two cases. One, of course, in that DDT was withdrawn from use through the legislative process, and it was a first case of its kind to undergo this ordeal by fire. The year was 1972, and the EPA had only just been formed. The second is that DDT, an early pesticide, was apparently effective with a broad range of insects and was used on virtually every crop grown. At first, the idea of eliminating DDT had farmers perplexed as to what could possibly take its place. Alar, on the other hand, was produced by only one company, Uniroyal. This allowed the company to make the necessary corporate decision unilaterally, without working through the tortuous federal EPA process. Also, while DDT was ubiquitous, Alar was used discretely on a few crops—mainly apples—though these crops did have wide distribution in the United States.

Finally, the evidence against DDT was drawn from its effect upon wildlife such as brown pelicans, whose depleted, thinned-out eggshells were heralding an extinction of future generations. But still, the concern at the time was for the health of wildlife. In the public consciousness, the threat to wildlife and its habitat was only remotely connected to its own health and well-being. People dwelling in cities and suburbs for the most part seldom viewed nature and wildlife, in any form, except possibly as intruders: the deer that stumbles through the garden, or the raccoon invading garbage cans.

The lobby against the continued use of DDT was therefore represented by naturalists, scientists, some old-line conservation organizations in existence since the late 1880s, and a new breed of environmental lawyers. The general public had a less active role in the battle for the banning of DDT.

In contrast, the presence of a potential cancer-causing substance, such as Alar, will always catch the public's attention. That was certainly true in 1959, when the cranberry became persona non grata due to a story, leaked to the press, that a cancer-causing substance had been used in cranberry growing. To this day, some Pine Barrens farmers contend the leak, and its timing just before Thanksgiving, the high point of the cranberry buying season, was politically motivated. This was a particularly painful experience for farmers because some states, such as New Jersey, were prohibited from using the sample herbicide by the Commissioner of Agriculture. Even though most farmers did not use the sample pesticide and, at that time, regulations on the federal level which looked into the health effects were nonexistent, people responded to the cancer-causing threat by avoiding cranberries for several years. The economic effects upon cranberry growers proved devastating.

In the 1989 story of Alar, we can see how much economic considerations, and the increasing awareness of environmentally related health problems by the public, led to a swift conclusion. Pressure from Mothers and Others for Pesticide Limits was certainly one key factor in forcing the issue. Not only did they lobby against the use of Alar, but they argued for important reforms in the way our government regulates pesticides and demanded some basic changes in our farming practices. They advocated that the EPA be required to take pesticides off the market if they were found to pose serious risks and that the FDA must improve its methods for sampling food and detecting pesticide residues. Finally, they asked that the federal government provide farmers with financial assistance and other incentives to switch to safer, low-pesticide farming techniques. These techniques include organic farming and integrated pest management. It was a tall order for our government to fill. Mothers and Others Against Pesticides was a project grown as a branch of the Natural Resources Defense Council. This was the same organization to publish the study on pesticides and food and implicate the dangers of pesticide to children. Mothers and Others went further in their campaign and organized town meetings nationally, particularly in the southern U.S., to spread their ideas. Lorraine Vole had been a community organizer for a national presidential campaign before joining the outreach staff. The organization sought and received visibility and legitimacy, respectively, from a Hollywood actress, Meryl Streep, as Chair, and a member of the Rockefeller family interested in environmental concerns, Wendy Gordon Rockefeller, as Vice-Chair. Their focus, initially was the Alar and Apples campaign. This effort proved to be largely successful. The result was enormous economic losses for the apple industry; it was also considered a tremendous victory for public health.

ECONOMICS AND FOOD—THEY GO TOGETHER

It wasn't the apple grower alone who stood to lose income from the Alar and apples scare. The effect would be felt in supermarket chains as well. Some supermarket chains, five in the United States and Canada, and a food distributor decided to take aggressive action to instill confidence in the buying public. This group announced in early fall of 1989 that it was taking efforts to reduce the levels of toxic pesticides on fruits and vegetables. It would do this by asking food suppliers to disclose all pesticides used to grow fruits and vegetables and would encourage growers to phase out, by 1995, the use of 64 pesticides considered potential carcinogens by the EPA.

The response to this effort against pesticide use was not uniformly praised. In fact, it was even considered alarmist by food industry trade associations and the EPA. In response, executives of the supermarket chains said they were taking the action because the environmental agency was moving too slowly in removing hazardous pesticides from the market.[7] It had become obvious to any one paying attention that public avoidance of certain food purchases could have dire economic consequences for supermarkets as well as farmers. By fall of 1989, a number of specialty supermarkets in the United States were setting aside counters for "organically grown food." Meanwhile, the supermarket trade publications were carrying articles about the problems of labeling something "organic" or "natural." How would they know? How could they know? Certainly, business was not willing to lose the public's confidence, but the labels had become meaningless. One article printed in Supermarket News states: "The rising interest in organic produce and an increase in sales has stepped up the push to forge a uniform, national definition of the category."[8]

Hoping to unravel the patchwork quilt of state laws and regulations, the United Fresh Fruit and Vegetable Association invited more than 60 growers, retailers, state and federal officials, and members of Congress to attend a meeting in Dallas to develop a framework definition for this emerging category. Thirty private certification groups and 17 state governments had already patched together a dizzying array of laws, regulations, guidelines, and standards governing organic produce. Other states planned to follow with their own rules, and Congress, too, had entered the fray.

The possibility of fraud in this respect and the economically dangerous possibility of the loss of public confidence in the ethics of food distribution were at stake. The food industry knew they had a reputation to maintain. For example, the California Department of Health Services shut down San Diego-based Pacific Organics for importing carrots grown in Mexico with the use of chemicals and marketing them in the United States as "organically grown."[9]

At the present time, 17 state governments have developed a dizzying array of laws, regulations, guidelines and standards governing organic produce.

The states that have adopted organic labeling regulations are California, Colorado, Connecticut, Iowa, Maine, Massachusetts, Minnesota, Montana, Nebraska, New Hampshire, North Dakota, Ohio, Oregon, South Dakota, Texas, Washington, and Wisconsin.

But while labeling regulations offer some definition, lack of national definition has interfered with interstate commerce. For example, California organic produce grown for one year in soil free of synthetic chemicals can't be sold in Minnesota, which has a three-year standard. Without knowing it, retailers can expose themselves to penalties for falsely advertising something not in accordance with applicable state laws and regulations. There is also a certification program sponsored or financed by a few states, including Minnesota, New Hampshire, Texas, Vermont, and Washington. Certification resolves the problem of false advertising, but is far from problem free. Lack of verification and varying state rules add to the difficulties. Nevertheless, the food industry now knows that the problem generated by lack of public confidence will simply not vanish. The Center for Produce Quality, formed only a year before by the Produce Marketing Association and the United Fresh Fruit and Vegetable Association, began a newspaper advertising campaign which stated, "The major concerns about fresh fruits and vegetables are deeply held and longstanding. They won't go away overnight."[10]

Mainstream citizens have become an active and vocal component of the environmental movement. Consumers can and do influence the marketplace in ways they could not have imagined. Knowledge of the potential threats to their family's health can shake Americans out of their apathy and affect their buying patterns, or even spur them to enlist in national or local community groups. The participation of mainstream citizens in the decision-making process of our country is a positive force to be reckoned with. The continuous participation of the public is vital to ensure the health of its citizens.

Critics of this upheaval in growth and management of food continue to insist that we are making "much ado about nothing." Their argument is that we are scaring ourselves to death, and that the media is largely responsible for this mishap. It is fairly obvious that the media does play a significant role in the transmission of bad news about food and water, and the environment in general. Just how responsible the media is in presenting these issues is a subject worthy of examination.

CHAPTER EIGHT

ARE WE SCARING OURSELVES TO DEATH?

The Media and the Scientist Report on the Environment

When we, the public, learn of an environmental crisis, that education and information is gained almost exclusively from the media. This includes print journalism and television news programs, such as the evening news, "The MacNeil/Lehrer Report," and "60 Minutes." For a more in-depth analysis a reader can spend hours pouring over highly respected science-related publications such as *Scientific American* or *The Ecologist*.

But since most magazines are monthly, the immediacy of a new, groundbreaking issue can only be captured by the daily press. Given the short deadlines for such stories, plus the fact that very few journalists are experts or even have the opportunity to specialize in areas of science, medicine, the environment, or health, the resulting public alarm over many environmental concerns is often blamed on journalistic sensationalism and inaccurate reporting.

As the communicator of bad news, are the media being blamed for the events that generate the news? Are they in fact much like the ancient Greek messenger, who by tradition was killed when he delivered unwelcome news, almost as if he had created the unhappy event by the very act of reporting it? We must now ask ourselves some pithy questions: Is the reporting of environmental news inaccurate? Is it deliberately or unintentionally sensationalized? Can the reporters themselves be held responsible for the subsequent public alarm and outcry that follow their disclosures?

These questions are vitally important. History has shown, in the recent Alar and apple scare and in the Amitrole and cranberry scare of the late 1950s, that public alarm and outcry can result in dramatic actions. In both cases, the public was informed of a potential cancer risk from ingesting certain chemicals. The report of this risk was first released through the media. And it was the media who provided the public with statements from scientific experts and governmental officials in an effort to communicate the complexity of the issue.

In the earlier Amitrole incident, the public learned of the health risk—the potentiality of thyroid cancer—from the print news; in the later Alar alarm, the public was alerted through a special CBS "60 Minutes" television program. The public, in both cases responded by boycotting the questionable fruit in the marketplace. The economic consequences were swift. The political consequences soon followed, with demands for a change in the way that the federal government, and particularly the much criticized EPA, protected the public's health. Thus the publishing of information on environmental health risks can have widespread effects and is not to be taken lightly. To rephrase our original question: are the media alarming the public unnecessarily by inaccurately reporting stories that ignite and distort the real situation? Or are the media simply performing a public service by reporting what the "experts"—the scientists—tell them? And, are all scientists equally capable of answering these questions? If not for the media, how else would people know if there were risks to their health and well-being? Some would argue that the nation's health and the state of the environment is already under expert control between the leadership of industry, scientists, and government, and that the American public is unnecessarily disturbed by these alarmist reports.[1,2] Not surprisingly, there are supporters and detractors for both points of view.

THE MEDIA CRITICS

As a critic of the media in this respect, Hodding Carter III, a political commentator who heads a television production firm, offers his viewpoint in the *Wall Street Journal*:

> Somewhere in the middle of the great Alar scare, a spokesman for one of the supermarket chains, stampeded into taking a brand of supposedly contaminated apple juice off its shelves, put the matter—and much of modern life—into proper perspective: "We're dealing with perceptions here," said the spokesman. "We're not dealing with reality."

To this observation, Carter noted the cynical Washington cliché: "Perceptions are reality."[3]

Carter offers the Alar case as a perfect example of reality falling away to perceptions. Following the Alar case study presented on "60 Minutes," Carter states that "all hell broke loose with schools dropping apples and apple juice from their menus, and supermarkets whisking products off their shelves. Parents by the thousands swamped Washington with protests." Carter further states:

> a public advocacy group, the Natural Resources Defense Council, used and was used by the CBS TV show to make the case that Alar, an agricultural chemical applied to a small portion of American red apple production, posed an intolerable threat to preschool children.

But, Carter concludes:

> the only problem was, and is, that a solid scientific case has yet to be made against Alar. Even if one had been made, the vast majority of all apples and apple derivatives have never been touched by Alar. And even those that were supposedly "contaminated" didn't rise to the risk levels detailed in the worst-case scenarios. To all of which the media now say, in effect, "Oops, so sorry. We only report the news. We don't make it!"

Finally, in this highly critical *Wall Street Journal* "Viewpoint" (an Op Ed piece), Hodding Carter III, summarizes a *Washington Post* report. The *Post* states, he remarks, that "a complicated scientific issue came close to being decided not by 'solid evidence' but by a frightened public acting on incomplete and often erroneous press reports."

When Carter wrote this piece in April of 1989, he could not have predicted the outcome of this Alar affair. He could not have known that the EPA would move toward the process of banning Alar from further use, or that his information would now be viewed as incorrect. As the EPA shifted position, the federal agency raised the estimated number of crops exposed to Alar from a low of 5 percent to a more substantial figure of 15 percent. His assertion, which formed the basis of his argument, that very few apple crops were exposed to Alar in the first place, no longer reflected reality. But the available information at the time of his article supported his statement that the press was overreacting. So it could be said that the press had overreacted, considering the existing knowledge at the time. Even if the press was alarmist considering what was known at the time—could the fact that the media generated so much interest have had something to do with the fact that the issue was investigated rather than dropped? In that case it can be viewed as a positive force. Later, it was found that the EPA had underestimated the figures. What had actually occurred was the following: The Alar situation was filled with conflicting reports.

The EPA initially quoted a low estimate on the number of crops exposed to Alar, and media people who followed the EPA report considered the ensuing coverage alarmist by comparison. On the other hand, journalists using the Natural Resources Defense Council report as their source of information quoted some disquieting conclusions and much higher estimates of the potential danger caused by Alar. In the latter case, journalists presented a far more threatening situation to the public. Furthermore, Hodding Carter could not have predicted that the apple growers themselves would voluntarily withdraw the use of Alar, nor that the fears for the 1989 apple crop without this growth promoting chemical were largely unfounded. Only later, during harvest season, did the public learn that the New York State apple crop, without the help of Alar, weighed in at about 990 million pounds in 1989—a 9-percent increase over the 910 million pound crop in 1988 which had been sprayed with Alar.[4] A *New York Times* article printed in mid-October harvest season offered us a surprising headline, "Alar Scare Unfounded, Apple Growers Say." What the article told us was that the fear of growing apples without the assistance of Alar was largely unfounded once the actual crop was harvested. The *Times* article explains that

> the amount of apples that drop from trees is crucial in the success of an apple crop. Once an apple falls to the ground, it cannot be sold as fresh and it is the fresh ones that command top prices. Alar was used since 1968 by growers to keep apples on trees longer. But large doses of Alar were found to cause cancer in animals and most farmers in the face of consumer complaints and an impending Federal ban, abandoned its use this year.[5]

So, we may ask: what is perception and what is reality? And the answer appears to be, a combination of both. In the case of the Alar "scare," articles which tended to minimize the potential health threat of a cancer-causing chemical gave way to an acknowledgment of risks as the federal government took concrete steps to ban Alar's use. So the banning of a pesticide becomes the evidence of potential harm to the public health. Yet, the banning efforts were preceded by weeks, even months, of denials that such a threat existed—hence, the complex workings of perception and reality. Once action is taken on the banning of a chemical, that chemical can no longer be considered harmless, nor can the press be called irresponsible for reporting that information. If the chemical were not dangerous then why would the government or, in recent cases, the food industry themselves, take corrective action. The answer to this query in some quarters is that public alarm awakened through the media is the true instigator of banning, and not the risk of the chemical to public health. How can we know what is the reality of any given situation when all of these forces come into play?

Here is where the media does play an important role, though. Not only do they communicate information, the information that they choose to publish takes on the mantle of legitimacy when it is printed or broadcast. In this sense, the perception of a few becomes the reality for many. That doesn't mean it is accurate.

If we turn back the clock to April of 1989, the press and the Natural Resources Defense Council study on pesticides and food were both sharply criticized. Disapproval was voiced by the EPA and by industry-backed research groups, such as the American Council on Science and Health, for inaccuracy and sensationalism. The assertion that Alar can lead to the risk of cancer was considered doubtful by industry and government; nevertheless, six months later, we find a *New York Times* story affirming that "large doses of Alar were found to cause cancer in animals," and that a Federal ban on its use was impending. Were these events the result of public pressure and alarm, or was there a scientific basis for banning Alar which had simply not been acknowledged before? If the former part of the statement is true, could farmers have lost money and time needlessly from an alarmist scare? If the latter part of the statement is correct, then we could be dealing with a serious societal problem. That problem, simply stated, is the effectiveness of government's role as guardian of the public well-being and the EPA's mandate as a federal protection agency. Our concern, then, is not only with inaccurate and alarming reporting, nor a galvanized public, but the dilemma of acknowledging publicly what is conceivably already known to a smaller circle of scientists.

The key to the dilemma lies with government leadership and scientists in the EPA and, perhaps, the FDA as well. Do they have access to information which would shed the necessary light on the question of pesticide dangers—or have the years of neglect in research on health effects left them, as has been noted before, without the data, time, and funding necessary to address the problem? Then the media have served as the bearer of information, inaccurate or not (we would, of course, prefer accurate), and the public as the source of political pressure. This at least gives us an important clue as to how our society works when it comes to risks, and to the process that creates the demand for clean food and water. One thing is certain, the Alar–Apple case led to anxiety within the food industry. Media reporting and public reaction in ensuing revelations of pesticide risks also led to comparatively brisk actions from the EPA. It is this proven and powerful combination that we must learn to apply more often in order to ensure the safety of public health.

PROPOSED BANNING OF EBDC

Certainly, the rapid announcement to partially ban the fungicide EDBC is an example of anticipatory responses to potential public outcry. In early December of 1989, less than eight months after the proposed Alar banning, the EPA proposed the partial banning of a class of chemicals widely applied to fruits and vegetables to control fungal diseases. This ban (in fact, a restriction on its use) was proposed because of concerns that this class of chemicals may cause cancer. The EPA said it was taking the action to allay "unnecessary public hysteria," an oblique reference to the removal of Alar the preceding spring.[6]

The fungicide, known by the initials EBDC, for ethylene bisdithiocarbamate, is applied to 55 fruit and vegetable crops, including apples, apricots, bananas, barley, broccoli, brussels sprouts, cabbage, cantaloupe, and carrots. We can recognize in the list below that nearly all the 55 fruits and vegetables are recommended by national cancer, heart, and diabetes associations. These fruits and vegetables are known for lowering blood pressure, decreasing cholesterol in our arteries, the control of diabetes through low fat diets, and controlling cancers, such as colon cancer, through the addition of high fiber foods (which comprise most of the list.) Yet, every food on this list is subjected to EBDCs, a class of chemicals used to control fungal diseases, and which may themselves cause cancer. And the list is quite long. Thus the EPA proposes a ban on the use of EBDCs for the following:

apples	eggplant
apricots	endive
bananas	fennel
barley	field corn
broccoli	green and dried beans
brussels sprouts	honeydew melons
cabbage	kale
cantaloupe	kohlrabi
carrots	lettuce
casaba melons	lima beans
cauliflower	mustard greens
celery	nectarines
collards	oats
cotton	papayas
crabapples	peaches
crenshaw melons	pears
cucumbers	pecans

peppers	rye
pineapples	spinach
potatoes	squash
pumpkins	**tomatoes**
quinces	turnips
rhubarb	watermelons

Almonds, asparagus, cranberries, figs, grapes, onions, peanuts, sugar beets, sweet corn, and wheat will not be included in the proposed ban of the EBDC fungicides. The three foods listed in bold were later taken off the list for proposed banning.

With such an extensive list of foods, there is no way to consider a boycott at the market place. Moreover, EBDC has been used for 55 years, since 1935. If there were any dangers present from the spraying of EBDCs, wouldn't we have known about them by now? Not necessarily. For one, cancer can occur after long periods of exposure; and, even more important, their epidemiological effects have not been well documented. A scientist with the New Jersey Department of Environmental Protection pointed out in an interview that unlike Alar, which is systemic (by becoming part of the plant fiber) and more serious in terms of health effects, this second group, the EBDCs, is less potentially harmful. We would hope that to be the case, for some of us appear to have been exposed to this class of chemicals for a lifetime, that is, those of us under 55. Nevertheless, with such an extensive list of foods exposed to the fungicide, only the efforts of government and/or industry can intervene in changing the situation. In fact, the EPA has proposed a ban under the agency's review process that would not take effect until the spring of 1991, unless the manufacturers of the chemical act voluntarily, and the industry, which in this case consists of four basic manufacturers, appears to be cooperating. The EPA would knock out 45 of the 55 food uses, effective January 1, 1991, though existing stock of EDBCs would be allowed to be used up first. The reason for this suggested restricted ban is that the use of EDBC fungicides is so widespread, its presence in our food now exceeds the EPA formula for a one in a million acceptable risk level. At the present time, all 55 crops are legal in their registered use of EBDCs, but the risk numbers are not acceptable to the EPA. They are currently 4 in 10,000, which is a worst-case scenario. What this means is that there is a risk that 4 people in every 10,000 will develop cancer from exposure to foods grown with the application of the EBDC fungicide group. When the first 42 of the 55 foods listed above were proposed for fungicide banning, the EPA recalculated the risk numbers again, and found that they were still not good enough to fall within the acceptable risk level. It was then that additional crops were added to the proposed ban list: bananas, potatoes and tomatoes. It is with this reasoning that 45 foods will no longer carry the fungicides, including the chemicals maneb, mancozeb, and metiram, used on about $12 billion worth of crops. But, 10 crops will continue to be exposed to the chemicals since, it is argued by the EPA, the estimated cancer risk from the use of EBDCs on those 10 crops is only 3 in 1,000,000 during a 70-year lifetime. This risk assessment falls well within the EPA's acceptable risk level, though it appears to be of little comfort to environmental and public interest groups who criticize the agency for failing to ban all uses of the fungicides immediately. But, as we already know, the EPA banning process is slow and tortuous. The leading manufacturers of the fungicides, anticipating a well-orchestrated public outcry, voluntarily removed 50 percent from permitted use on crops, and made this effective by January of 1990. By contrast, with the EPA moving toward a partial banning, it will be the spring of 1991 before the proposal is finalized. Still, when all is said and done, there are 10 crops which are not affected by the proposed ban. Those are: almonds, asparagus, cranberries, figs, grapes, onions, peanuts, sugar beets, sweet corn, and wheat. These crops will continue to be grown with the use of EBDC fungicides.

Though the latest pesticide alarm affects a wide range of foods—namely, every fruit and vegetable in the green-grocer section—the announcement of this concern was given far less long-term coverage by the media than was the Alar case. We could conjecture that since the industry and the EPA were anticipating public outrage, they acted swiftly to demonstrate their concern for public confidence. In doing so, the media had less of a role to play in "heating up" the situation. If this assumption is correct, then the media have taken on a powerful role with respect to these public health issues. They are not only reporting the news, they are defining it as well. Are the media then instigating public alarm—and are they doing so unnecessarily? Was Hodding Carter III correct when he lashed out at what he considered inaccurate and irresponsible reporting?

ARE WE SCARING OURSELVES TO DEATH?

In an effort to tackle this compelling issue, The Center for Communication, in New York, an educational organization devoted to exploring the role of the media in society, put together a seminar titled: "Reporting on the Environment: Are We Scaring Ourselves to Death?" In a piece written by the Center's program director, Irina Posner, and executive director, Catherine Gay, they ask: "How does the press report on the environment? Since the way information is presented is often as important as its content—especially with issues dealing with risk—the role of the media is central to the creation of social policy."[7]

The Center selected six experts to assess the role of the media in developing social policy: Jonathan Piel, editor of *Scientific American*; Edward G. Remmers, a scientist and vice president of the American Council on Safety and Health, an industry supported, science-based organization highly critical of the Natural Resources Defense Council in general and of the Alar report in particular; Michael Greenberg, a Rutgers University professor of public health who is highly critical of the press; Gary Soucie, executive editor of *Audubon*, and, Laurie Garrett of *Newsday*. I served as moderator.

While Michael Greenberg and Edward G. Remmers provided criticism of media reporting on the environment, their arguments emanated from quite different perspectives. Remmers, as vice president of the American Council on Science and Health (ACSH), asserted that the most important public health concerns for Americans are cigarette smoking, AIDS, drug abuse, all of which are problems determined to some extent by individual behavior. On the other hand, Remmers defines environmental issues, such as PCBs, EDB, dioxin, artificial sweeteners, and irradiated food, as simply "hypothetical" health risks. Scientific ignorance, Remmers states, goes a long way in creating those misunderstandings.

> I believe that the scientific illiteracy of the American public is another cause for serious concern . . . unless our educational system is improved, Americans will continue to be misled and misinformed about scientific issues, poor-quality science will continue to be used by our regulatory agencies, and we will not be able to maintain our competitiveness with other key industrial nations.[8]

What the ACSH aims for is to be looked upon as an educational organization to whom the press can turn for accurate, "mainstream" science. But if it is industry backed, can it be totally "mainstream" in terms of objectivity? This largely industry-backed organization prefers a stronger voice in the media when it comes to discussing scientifically related issues.

Much in support of the American Council on Science and Health in respect to scientific illiteracy is Dr. Manfred Kroger, a professor of food science at Pennsylvania State University. "The general public is grossly illiterate in matters of science and technology," Professor Kroger states.[9]

> The media will filter and condense all the information they choose to disseminate. Even with major stories, we are talking seconds rather than minutes. . . . So far, scientists have failed as professionals in communicating; and judging by the nonsense that people read and hear and believe about food, we food professionals also have little to be proud about.[10]

While few could disagree with Kroger's concern for the clear communication of scientifically based issues, his emphasis upon the authority of the scientist versus the right to correct science and technology information leads to some questions about his view of that relationship. For example, he comments that

> the vast majority of the public is illiterate when it comes to science and technology. Not only that, there are forces of

irrationality, antiscience, and antitechnology in this world. There is tremendous preoccupation in our society with astrology, ESP, psychic phenomena, witchcraft, superstitions, demonology. . . . We in the scientific establishment must use all the ploys and plays available to broadcast our message.[11]

But precisely what is that message? More than in any other field, the food scientist tells us,

> the food area is subject to horrendous misinformation . . . that food quackery runs wild and cuts across all social strata . . . that there are fraudulent potions for sale . . . or undefinable so-called organic foods; . . . there are unwarranted fears, such as of water fluoridation, food additives, food irradiation . . . the whole field of dietary advice has largely passed out of the hands of nutritionists into the fingers of entertainers. . . .[12]

All in all, Dr. Kroger concludes that the wave of health consciousness sweeping this country is "accompanied by irrational calls for super-health foods and a super-clean environment."[13] Once having established the dangers of a misinformed public, Kroger goes on to offer advice to scientists who choose to educate the public and undertake the awesome task of media exposure. Thus he advises:

> • Control the environment, manipulate the setting. Insist on the best camera angle. Direct your answer into the lens.
> • Use props, use alliteration, by all means use humor, use ploys. (A guaranteed winner is: "I shouldn't tell you this but . . ." It will be on the air at eleven. The press will quote you.)
> • Answer questions with the iceberg technique: The tip only, details don't make headlines.
> • Deflect and ignore questions you don't want to touch.
> • Don't let the audience control the show by asking questions.[14]

At this point, we can recognize that media skills adopted by our politicians and actors have now entered far-reaching sectors of otherwise remote professions. Further, while we find in this outspoken scientist a strong emphasis upon educating the public about science, there is a simultaneous message. This second theme tells us that the "mainstream" scientist is the only responsible authority on the subject of food and food safety.

Dr. Elizabeth M. Whelan, executive director of the American Council on Science and Health, to which Remmers is vice president, is even more outspoken in her assault on the public's reaction to environmental reporting. An epidemiologist, Whelan wrote in the *Los Angeles Times* that,

> what this country needs right now is a national psychiatrist to determine why Americans are discarding wholesome food in a mindless effort to reduce the risk of cancer. . . . Nosophobia is defined as a morbid dread of illness. It is a form of psychosis that is extremely hazardous to both physical and mental health; its effects on our nation's children are particularly insidious . . .
>
> The most recent and to date most virulent episode of national nosophobia was triggered by the release last month of a report by the Natural Resources Defense Council which opined that trace levels of agricultural chemicals cause cancer in kids. The sole basis for this phenomenal and frightening charge was the assumption that huge levels of chemical pesticides that cause cancer in mice must also be carcinogens in humans . . . [15]

Without having to interpret the message in this one article, we learn that the American public is close to psychotic when it comes to the fear of illness, and that cancer research on mice is hardly conclusive. If the public becomes confused from these messages it is because the article is both accusatory, and inaccurate. Cancer research is normally conducted on mice as laboratory animals that reproduce rapidly into third and fourth generations, and from which scientists can extrapolate information on possible human carcinogens.

Compelling evidence, however, can be drawn from epidemiological studies. This public health research methodology includes the gathering of health histories and statistics on groups of people, communities, or even total societies in order to document the prevalence of illness and death; without data collection this type of statistical research is not possible. It's puzzling that Dr. Whelan uses this everyday scientific procedure of cancer research using mice as an example of what "triggered national nosophobia," leading the reader to conclude that animal research in cancer is a deviation from what is normal procedure. Such assertions are even more puzzling to the public when the results of these animal research tests are reported by others in the print media to be particularly disquieting. We read from a *Nutrition Watch* journalist, Mary Painter, that EPA's acting deputy administrator, Jack Moore, "called the company's (Uniroyal) own data on the chemical (Alar) 'most troubling' adding that it showed that high doses fed to mice produced 'life-threatening tumors.'"[16] Thus, tumors in mice resulting from a chemical alert scientists to potential cancer threats to humans.

ENVIRONMENTAL RISKS AND
THE EVENING NEWS

Professor Michael Greenberg, with colleagues David B. Sachsman, Peter M. Sandman, and Kandice L. Salomone of Rutgers University conducted research on "Network Evening News Coverage of Environmental Risk."[17]

By examining the evening news broadcasts of the major networks—ABC, CBS, and NBC, though not cable news—the authors note that carefully crafted and expensively produced evening news broadcasts devoted 1.7 percent of their air time to 564 stories about man-made environmental risks during the period from January 1984 to February 1986. However, they found little relationship between the amount of coverage in the news and the corresponding public health risk. A potential health risk was not the key reason for coverage in the news, at least according to these researchers. Instead, the networks appeared to be using traditional journalistic criteria of news. Such determinants as timeliness, proximity to the viewer, consequences, and human interest were the major reasons for focusing on an issue as news. Furthermore, Greenberg concludes that the broadcast criterion of "visual impact" was perhaps the most significant factor in determining the degree of coverage of risk issues. Visual impact literally refers to dramatically photographic subject matter. In other words, a story on lung cancer would provide fewer visual images and less impact than an earthquake, which has the dramatic visual impact of fallen buildings and victims. This visual impact, according to critic Greenberg, determines the amount of news coverage on television network news, not the more important question of potential or serious risk presented to the public.

WHAT ARE THE SOLUTIONS?

The solution, according to this Rutgers research group, is for the media to consult scientists more frequently. While that is certainly a rational suggestion, we may also have noticed, as has the media, that scientists are not all of one mind. There are as many varying opinions on the potential health risk of Alar, or DDT, or EBDCs as there are variations in the risk assessment itself. Many scientists cannot offer us an objective or "value-free" assessment, for they are either working for organizations seeking to minimize the problem of "public hysteria," or they may genuinely believe that the costs of public health risks have to be balanced against the benefits of technological advances. Both of these approaches, of course, need to be defined. Is the "slight" risk of contracting cancer from Alar, for example, too high a price to pay for crisper apples? What is troubling about such a hypothesis is the assumption that the use of Alar is the only means to achieve crisper apples; or that the chemical in itself is a positive result of technological advancement. We do tend to view technological change as positive, even though that change may not advance or benefit our society in the least. Yet we see attempts to justify and continue the use of certain technologies despite the dangers they pose. We forget that some solutions found in earlier times were discarded altogether, though they were not necessarily ineffective but just replaced by alternatives. After all, insect control is not a recent practice. The ancient Greeks and Romans used sulfur to rid their crops of insect pests, and the Chinese used arsenic sulfide to control many plant-eating insects long before the watershed development of synthetic insecticides. As recently as one century ago, U.S. farmers were using natural poisons such as arsenic, lead, and copper on fruit crops with little outcry from the public, though European importers voiced some reservations about the safety of American produce. In fact, some of these natural poisons are still employed in restricted agricultural use. So the problem of finding effective plant protection is ancient. And the problem of making this protection effective without disturbing the health and safety of people and the environment is certainly not a new concern. But when we turned to new chemical technologies some 50 years ago, we believed that we had found a major breakthrough in agriculture, and for a while it certainly appeared to be the case. These earlier-used substances could have been detrimental to public health, and quite possibly abandoned when more effective though not less harmful practices were developed. The question we raise now is whether our current practices are more effective and less harmful to our health. If not, then how can we justify using them.

In a particularly vituperative article in the *New Republic*, Henry Fairlie, an armchair critic of the American scene and himself a British expatriate, comments that

> the idea that our individual lives and the nation's life can and should be risk-free has grown to be an obsession, driven far and deep into American attitudes. Indeed, the desire for a risk-free society is one of the most debilitating influences in America today, progressively enfeebling the economy with a mass of safety regulations and a widespread fear of liability rulings, and threatening to create an unbuoyant and uninventive society.[18]

Fairlie sees this point of view as another distortion of the philosophy of rights underlying the Constitution,

> as if the Declaration of Independence had been rewritten to include freedom from risk among the self-evident rights to life, liberty and the pursuit of happiness. This morbid aversion to risk calls into question how Americans now envision the destiny of their country.[19]

Peter Passell, in the *New York Times*, says that according to psychologists, one way to restore a sense of balance between risk and reward is to package information about risk in ways that people can more easily grasp. But economists and legal scholars argue that equilibrium will not be possible until there is wider understanding that unfamiliar and often unmeasurable risk is the price of prosperity and societal achievement.[20]

But the idea of "unmeasurable" risk is precisely part of the public's apprehension. To make matters worse, once having established that these risks are often unmeasurable, Passell goes on to offer statistics to demonstrate that Americans have never been safer. (Apparently there are no measurements on risks, but ample numbers to prove safety. Unless logic fails us altogether, the second part of the statement is not possible without first including the effects of risks.) The reporter goes on to say that

> life expectancy at birth in 1986 was 74.8 years, up a full four years since 1970, largely because of a dramatic drop in deaths from heart disease and strokes. Cancer deaths, adjusted for age, are up slightly in the last two decades, but not, apparently, due to man-made hazards in the air, water, workplace and food supply.[21]

Again, we are faced with a problem of logic. If there is limited information on the risks of man-made hazards in the air, water, workplace, and food supply, how then can we conclude that these same causes were not the contributors to the rise in cancer deaths. Without measurable information, we cannot conclude either way. We can merely state that we don't have the information necessary to accept a cause–effect relationship between these man-made hazards, and the rise in cancer deaths. At this point, the informational handicap—lacking measurable risks—is a problem when we attempt to forge an understanding of the effects of pesticide use on society. The gap in information stems from decades of not gathering knowledge and statistics on health effects. For 30, nearly 40 years, pesticides were registered for use only on the condition that they perform as the chemical company claimed—that they have the capability to destroy the targeted pest. Registration of chemicals by the federal government had no requirements for documenting health effects. Changes recently have been suggested for the FIFRA, the Federal Insecticide, Fungicide, and Rodenticide Act, and the EPA has its role as watchdog of public health and interest, but for the most part the question of public health risks leaves us with a lot of blank pages. Hard evidence is only just coming to light; unfortunately, it lacks the historical record which could serve as the basis of epidemiological studies.

As to the responsibility and reliability of the press in reporting on these environmental risk issues, it is fair to say that they alone cannot be held responsible for the public outcry on potential problems. Real concerns do exist and require addressing, however late in time. What the media could improve is in the drafting of more balanced articles and programs giving the views of varied scientists and public officials. Since we already acknowledge that scientists are seldom "value free" themselves, it would help enormously to present the many sides of an issue before coming to rapid fire conclusions. But the daily media has deadlines—nothing as dead as yesterday's newspaper, the expression goes. Often deadlines deny them the luxury of time to explore an issue, as a scientist in research is afforded. What might improve their reporting of environmental concerns would be a greater awareness of differences in scientific disciplines. Dissimilar in focus, methodology, and viewpoint, varying fields of science explore questions unique to their own particular discipline. For example, a naturalist would be concerned with the fate of wildlife and head for the fields to conduct sampling and observations; the toxicologist would more likely be found in the laboratory investigating chemical reactions; a sociologist would enter the field of people for interviewing and observing, to compare statistical evidence on the effects of certain practices upon the community or the nation as a whole. Each of these endeavors could lead to very different assessments of a given situation. There are also microbiologists, virologists, botanists, and organic farmers directly involved in questions concerning agriculture and the environment. In other words, instead of merely increasing the number of sources consulted, it would bring greater balance to an analysis if entomologists, naturalists, biologists, public health officials, physicians, and social scientists such as sociologists, economists, and even historians were consulted. At the moment, only one or two disciplines are interviewed for confirmation of information. We can understand the problem of time shortages and deadlines for the journalist, but in-depth reporting is to everyone's interest. Indeed, what we need from the media are analytical articles and programs drawn from the expertise of a spectrum of experts which will allow the public to become educated while reaching for its own conclusions.

The pragmatic endorsement of the chemical age and the industries which produce that technology is facing public scrutiny as well. The public is less amenable to leaving all the choices and decisions to the figures in authority. A certain amount of trust has been eroded. Those scientific pragmatists who scoff at public demands for a pure environment and clean food and water offer instead a schema of a necessarily compromised natural world as the price paid for containing human hunger. But what answers do they give to the concern for threats to health and the environment? And is the question of hunger mutually exclusive from the fate of the environment? The answer of course is no. A drought caused by contamination of groundwater and depletion of the overall water supply in agricultural areas would certainly contribute to a cutback in food production if not a famine. Depletion of minerals in the soil and overforestation have changed the fate of forests and the landscapes on which people live. Dare we be so cavalier with ours simply because we have a storehouse of technological answers? Can we really fail to expect a clean environment—is it so unreasonable to seek alternatives to the massive dosages of chemicals currently in use in agriculture? Are these unwarrented fears based on superstition and ignorance, or are we simply discovering the depth of very real problems and therefore expressing appropriate alarm as life-threatening situations are belatedly uncovered.

Americans, we are told again and again in the popular press, are much too concerned about their health. This concern is an outgrowth of an overly affluent society, so we are told. But in fact the environment has become an international concern in industrial societies, postindustrial societies, and in developing countries. A wealthy nation can boast of many technological advances, and yet, some of those practices can be detrimental to the health of people, and to the world they inhabit. Is the demand for a clean environment unreasonable and unrealistic? Is the demand for clean food and water a basic right—or is it an impossible dream? How we address these questions will grant us a key into the future—a vision and forecast for the twenty-first century and beyond.

CHAPTER NINE

RETURN TO THE FUTURE

The 1990s is not simply a new decade, it represents a major turning point in our history. The decade heralds the turn of the century, and the new millenium: the twenty-first century is near at hand. We enter this new era with a sense of impending transformations, for as the 1990s progress, we conjure up all the mythic expectations surrounding our idea of the twenty-first century. What we most envision is change—vast change stemming from the relentless pursuit of technology, an application of science that will continue to alter significantly our concept of time and space, and the world in which we live. The burgeoning populations, scale of food production, towering architecture, modern forms of transportation are all products of the twentieth century. What lies ahead for the next millennium?

Let's go back in time, to the 1890s—that era in the last decade of the nineteenth century, which offered some dynamically changing visions for the future. Those latter-day Victorians accepted with apparent ease the rapid speed of change in their eagerness to greet the oncoming twentieth century. The nineteenth century had been fraught with problems resulting from the massive growth of industrialization. The aftereffects were seen in the rise of concentrated populations in cities and disastrous public health conditions in both factories and homes. The burning of coal, for example, blanketed cities with blackened soot, and dark clouds of its pollution continuously hovered over London and New York. From the ensuing damage, tuberculosis claimed the lives of the poor and rich alike at an astonishing rate. Determined to leave these problems in the past, the citizens of the nineteenth century directed their efforts and hopes toward a salvation that would be made possible through the miracles of science and technology. Optimistic thinkers predicted the arrival of a more perfect utopian world as the twentieth century ushered itself in. H. G. Wells, the British journalist and reformer, offered startling visions of the future, based on his early training in biology, in his highly popular fiction. In actuality his books were thinly veiled treatises with a common underlying theme: scientific progress would bring about the social millennium—the long awaited age of reason. Books such as *The Time Machine* (1895), *The Stolen Bacillus* (1895), *The War of the Worlds* (1898), and *The Invisible Man* (1897), all published during the last decade of the century, foretold of miraculous scientific discoveries which could bring forth a new and highly rational society—but only if human beings could resolve their destructive and aggressive tendencies with respect to the earth, and each other. Like his prototype author, Jules Verne, H. G. Wells offered his turn-of-the-century generation an optimistic, long-term view of the possibilities of science or, what we would today call, technology. Would technology eventually help our human societies to live in peace? Perhaps if technology could help us create adequate food supplies through a more efficient use of nature, there would be fewer reasons to create war as a solution to the problems of scarcity (if that in fact is a cause of war). This view permeated our zeitgeist throughout the twentieth-century, and history has shown that revolutionary discoveries have improved our quality of life. Yet, unanticipated, pernicious problems have also arisen from the application of some of these discoveries. "The Chemical Age," for example, has been a mixed blessing, seeming to solve some age-old problems while opening up a Pandora's box.

On the cusp of the twenty-first century, we continue to anticipate the miracles that technology will bring to improve our present way of life. This last decade is focused on the future, and on the impressive changes we anticipate will resolve some of our pressing twentieth-century problems. More and more we find ourselves becoming future oriented, as we contemplate the new advances that will a take place during the next hundred years. We tend to view the future as very near at hand, and like the latter-day Victorians, we look to the oncoming century to solve many of the dilemmas we have inadvertently created.

With the exception of Adam and Eve, who had no prior history to contend with, people have always inherited both the progress and problems of former generations. All the flaws, disasters, and advances of the previous generation are left for the children. Nineteenth-century air pollution problems resulting from widespread coal burning became our twentieth-century acid rain dilemma. By the same token, twentieth-century problems involving the effects of pesticides will become the twenty-first century's groundwater pollution problem and will jeopardize the safety of its citizens' food. So despite forecasts about forging brave new worlds with the astonishing power of technology, a good part of our planning for the future needs to be devoted to correcting the ills that we, as well as former generations, have neglected. These ills would form a litany far too lengthy to ever list in a single volume, particularly with regard to the environment. A condensed list, however, would include: urban smog, acid rain, toxic air, water pollution, depletion of the ozone layer, global warming, pollution, and decline of ocean life. The list could go on and on.

Environmental problems which pose a threat to public health, such as the presence of pesticides in food and water, for example, have been of particular concern to the populace. Pressure to resolve these problems has led to a complicated series of actions and reactions, spanning 55 years in this century.

Many of our environmental woes began with the introduction of chemicals. In the mid-1930s, the beginning of The Chemical Age dawned with the introduction of man-made pesticides into agriculture. The first group of synthetic, man-made chemicals, known as chlorinated hydrocarbons, provided the most widely used and direct means of killing and controlling insects and related pests. Since pests had been the plague of farmers for centuries, swift removal of the ancient demons of the fields was one more test of the modern miracles of the chemical age. Farmers and citizens alike welcomed the modern application of science which would make crop devastation a dim memory of centuries past. It was also viewed as the twentieth-century's incredible application of science that would open up a whole new world of conveniences.

The first synthetic insecticide developed, DDT, was initially used in 1942; since that time literally hundreds of other chemicals have been processed. After the introduction of DDT, a repertoire of chemicals, including herbicides, fungicides, miticides, and rodenticides, were developed to control all kinds of agricultural pests. When DDT was first introduced, society looked upon it as a boom to agriculture for its astounding ability to increase the yield of crops by depressing the ravages of insects. But 20 years down the road, during the early 1960s, the biologist and writer Rachel Carson alerted the public to its deleterious effects. She warned that DDT could interfere with the ecological chain of life; it could render certain wildlife obsolete; it could cause cancer in humans; and it could eventually result in resistant strains of insects. The medical and biological evidence to support these claims remained largely unrecorded by the government or the chemical industry for a surprisingly long period of time. When some indicators did surface, they were investigated through independent, nongovernmental studies conducted by scientists largely outside of government or industry.

In the latter 1960s, the effect of DDT on wildlife provided the bellwether for an environmental alarm. Another 10 years passed before DDT was officially banned in the United States after a fierce battle of conflicting forces among citizens, environmental organizations, government, agriculture, and industry. Despite this momentous decision, which closely followed the establishment of an entirely new federal agency, the EPA, mandated to protect the environment, hundreds of pesticides continued to be developed and used in U.S. agriculture and for overseas exportation. The burgeoning chemical industry was undaunted after the fall of the most powerful and widely used chlorinated hydrocarbons. American substitutes for DDT in the 1970s became the pesticides within the organophosphate and carbamate categories. Organophosphates in some categories can have a high toxicity for both humans and the environment, but at the time they were considered less harmful than chlorinated hydrocarbons, because unlike DDT they were thought to degrade in the environment. However, the use of any chemical that poisons, also has the power to kill. The poisoning of farm workers described in this book was the result of organophosphate poisoning from a combination of chemicals so potentially lethal and unstable that even heat and human perspiration in the fields could affect their properties. And these were the "safer" pesticides suggested in lieu of DDT.

The insecticidal properties of this second chemical category, the organophosphates, were discovered in the 1930s in Germany and were first used by that government as weapons of war against human beings. Some forms of these chemicals were used as nerve gases while others were relegated to insecticides. So the use of insecticides offered the potential to harm people in ways still not fully understood. Despite the banning of DDT in the early 1960s, Americans would continue to live in a world where food production was heavily tied to hazardous chemicals, right into the 1990s.

Today, the problem is anything but solved. The three chemical groups, organophosphates, carbamates, and now, to a lesser degree, the older chlorinated hydrocarbons, are used predominantly for the control of agricultural pests in a massively consolidated agribusiness system. Suburban Americans also make use of them in their homes and gardens. On rare occasions a hue and cry arises over a particular chemical that is reportedly harmful to human health. The proof, however, is not readily available since the extensive testing of pesticides for registration purposes has until quite recently only dealt with determining the pesticide's mode of action in the field. In other words, we have studied how a pesticide destroys the insect, plant, or weed of choice, but not its effects upon human health. To further compound the problem, throughout its long history of use, only the active ingredients—those ingredients which directly destroy the insect—have been tested. The inert ingredients, meaning those parts of the chemical formula which are present as emulsifiers or binders and do not directly affect the target pest, are simply ignored, and their potential harmful effects escape undetected. In terms of detecting physical symptoms, the most compelling evidence, of course, would be a history of the health effects of pesticides over time, particularly since cancer develops over long periods of time. Yet that kind of monitoring has simply not been done. Considering the hundreds of chemicals used in food production, we find that very few have been targeted for banning—on average approximately one per decade.

A few pesticides have caught the attention of the concerned public, as in the 1989 alarm over Alar. But that concern was generated as a result of the actions of independent environmental groups and organizations which exist outside the domain of government and industry. Such groups as the Environmental Defense Fund, or the Natural Resources Defense Council, or Greenpeace sponsor some research which can provide alternative views about the safety of our chemicals. Generating unrestricted research reports by these environmental organizations has the effect of overriding the role, and at times forcing the hand, of the EPA. Unfortunately, the federal agencies' positions in protecting the environment are sometimes tempered by the countervailing interests of industry. The

NRDC has subjective leanings as well, but its leanings toward the citizen may counterbalance the EPA's leanings toward keeping business happy, as well as protecting the environment. Forcing the hand of the EPA by publishing independent research reports sometimes results in actions that might not have been taken otherwise or, at least, might not have taken place so quickly. The Alar debate is certainly an example of this push–pull effect.

The media have also played a significant role in communicating the concerns of independent environmental organizations. When the public responds to the media's reports with alarm, the reactions are felt in the halls of Congress and in the marketplace, and then the marketplace usually responds very quickly. In fact, the public boycotting of certain foods with questionable health effects has proven to be a highly effective means of lubricating the machinery of change. For example, DDT as noted earlier was banned in the United States (but not for exportation) after 10 long years of debate and court action in 1972. By contrast, Alar was withdrawn from use by the apple growers during the same year as the alarm was sounded, in 1989. The major difference in the two cases was the targeting of a single food in the marketplace, specifically, apples. With only one or possibly two food products in question, the public's boycotting is entirely possible and potent. When the economy is affected, a move to change a system or alter a product can be impressively swift, particularly if there are only one or two food products involved. Industries do not survive long if their products are avoided by the public. The dollar talks fast. A rediscovered power in the marketplace offers the American public an opportunity to demand environmentally sound practices—once they know which products are harmful to their health. That information, unfortunately, has not always been readily available—either because the research and data do not actually exist, or because the knowledge does exist but has not been written in a style understood by or shared with the general public.

Some scientists complain that the public is scientifically illiterate and therefore not capable of dealing with technical information. But the public does understand threats to its health when the information is given to them in a lucid manner. More importantly, the public has a right to know. Unfortunately, the public is often given confusing and conflicting information. Scientists as spokespersons may sound uncommitted and equivocating on the subject of possible, or probable, causes of cancer, because their statements are usually drawn from predictive evidence. This evidence, gathered from research on rats, and rates of probabilities in risk assessments, is difficult to synthesize into a simple statement. All the public really wants to know is "will this kill me?" and "how can I stop this from happening?" Moreover, we may ask, should the public be required to assume the role of the constant watchdog over health and environmental practices at all times? That is a function it has somehow inherited, but must this continue? The answer is yes. It is important for government to play its part in protecting the environment and the people who live in it; since that role is often altered by the politics of the marketplace and the shifting agendas of varying political administrations, the public cannot afford to forfeit its own responsibility. The early years of the decade of the 1980s was certainly an unhappy example of the consequences of public disinterest in health and environmental issues.

Hence the American public needs to become a major player in its own protection. But is the public capable of understanding these issues? Some of the more vocal scientific experts contend that they are not. Eradicating the accusation by some scientists that the American public is "scientifically illiterate" would be a worthwhile endeavor. The more we learn the more we will understand. Educating ourselves and our children about scientific and environmental issues as well as the meaning of probabilities is essential in looking out for our best interests. Certainly we don't need doctorates in science to understand that specific chemicals in certain amounts can endanger our lives and that immediate action must be taken to correct these health hazards. The public needs to know what dangers exist, how dangerous they are, whether there are safer alternatives, and finally how to set the wheels of change in motion. Instead, the general public is usually exposed to an enormous amount of information through the media, but often these news programs are not sufficient in time or depth to present adequately these crucial stories. Special programming, such as documentaries which encourage critical analysis from the viewers, needs to be expanded and encouraged both at home and especially in the classroom. Granted, this type of programming may not be entertainment *per se*, but it is hard to think of oil spills or pesticide poisonings as entertaining. What we would hope to aim for in this type of programming is the presentation of interesting and compelling knowledge that will provoke the viewer to contemplate the subject at hand. The more we strive for entertainment when it comes to television news, the less the news will reflect reality in the public's mind. The media generates concern on important issues, especially through television. Television news particularly should employ people of some level of scientific expertise and objectivity to analyze pertinent scientific and environmental issues in depth.

In summary, then, the use of pesticides, their effects

upon food, water, and our health, is very much a twentieth-century problem. As we weather the 1990s, our search for answers is rapidly spilling over into the next century. Do all these thoughts suggest that we are optimistic about the future of our environment in the twenty-first century? Are we being idealistic when we talk of building a healthier environment with clean food and water as our natural right? There are critics of the environmental movement who insist that no advances are made without costs—no technological innovations are discovered without risks to be paid for, even at the cost of human lives. One social critic, Henry Fairlie, formerly of the *New Republic* (recently deceased) gave vent to his distaste for the purveyors of risk intolerance. He labelled the phenomenon "America's morbid aversion to risk."

AVERSIONS TO RISK—IS IT REALLY A FEAR OF LIVING?

Fairlie was a trenchant social critic who authored one particularly noteworthy article called "Fear of Living." In it, Fairlie refers to the irrationality of public attitudes toward risk. He portrays Americans as timorous beings, demonstrating a flaw in their character which renders them unwilling to take unnecessary risks. He comments,

> if America's new timorousness had prevailed among the Vikings, their ships with the bold prows but frail hulls would have been declared unseaworthy. The Norsemen would have stayed home and jogged. Columbus's three tubs would not have been allowed to sail . . .[1]

It is necessary to point out to Fairlie's proponents that the voyage of Columbus or even those of his predecessors, the undaunted Vikings, were not in themselves societal risks.

There were probably risks to the crew and its captain, but those hazards did not affect an entire society or country, their health and well-being, as might the use of chemicals in the food and water supply. The analogy used by Fairlie is untenable. Further on, the author states that the origins of this widespread refusal to accept a sometimes high level of risk as a normal and necessary hazard of life has its roots in the 1970s: "as Americans lost heart in the prosecution of the war in Vietnam . . . the vanguard of the 'Me Decade' turned to lavish care for the environment, the snail darter, and their own exquisite, often imagined, physical and emotional well-being."[2] (The snail darter is a tiny organism whose continued existence became the key argument for environmental groups campaigning against the flooding of a river, the snail darter's habitat, in order to construct a dam. The snail darter has often been used as an example of the excesses of environmentalists over an issue of change and risk.) Fairlie comments incorrectly that the risk-averse groups are drawn from the privileged class. In fact, concern for the environment, which began as an interest of the wealthy conservationists at the turn of the nineteenth century, is now very much of a mainstream American concern. There are currently 75 national environmental groups alone, and thousands of state and local-level groups, many of them focusing on the question of clean food and water.

Furthermore, the idea that risk is normal and often necessary to the continuance of our way of life is spurious as well. While a certain amount of risk is always involved in the employ of new technologies, we have not only created certain intolerable risks, but we have generated many which could be avoided. In many cases, we have been downright careless. We have been killing pests with chemicals, without giving thought to the effects of these practices upon health or the environment. We have applied heavy doses of pesticides without knowing what other organisms or life forms will be destroyed as well. We have had alternatives and options to which we have not given consideration. For example, farmers are now able to protect everything from apples to zucchini using biological control methodologies. Employing about 50 insect pests and by using one or more natural enemies, usually imported from the pest's homeland, a whole range of fruits and vegetables can be grown without the use of massive doses of chemicals. There are also beneficial microorganisms which can serve as living barriers between pests and crops, and as deterrents to infection or pest attack. We can create self-defense systems within the plant itself using genes produced from a strain of the insect pathogen. Sadly enough, through our irresponsible actions, we have created problems we haven't really tried to solve. These problems exist as an outgrowth of the changes in agricultural practices. These changes have had a tremendous effect upon the food, water, and health of this nation. The answer to solving many of these problems lies in how we handle the tenuous relationship between agriculture and the environment.

BALANCING AGRICULTURE AND
THE ENVIRONMENT

The growth and production of food is intrinsic to the survival of people, any people, regardless of the size of the populace or the method of agriculture used. As populations continue to grow in size, the demand for large-scale food production leads to a search for alternative, seemingly efficient agricultural practices. These practices may also have damaging effects upon the environment and in some cases this damage will be wreaked upon the very land and water necessary for that food production. Therefore, the method a society chooses to ensure its short-term survival will also threaten its continued existence. The answers we choose will affect the very balance of nature and the survival of future generations.

In less industrialized societies, slash-and-burn agriculture often destroys the ground cover and erodes soils. Deforestation and the clearing of rain forests, an activity conspicuous in Brazil, is yet another example of people depleting the soil by cutting down trees and clearing the land for farming. These new settlers open the land once protected by an enormous canopy of trees. The exposed land now becomes vulnerable to wind and erosion.

In the high-tech agricultural world of the United States and Europe, our methods may be more sophisticated, but no less deadly. Our heavy dependence on pesticides over the past 50 years has left a trail of problems, including groundwater pollution, and the possible detrimental health effects upon the human population.

Why do we continue to use these destructive practices, even after we have learned of some of their most damaging effects? There is no simple answer to this question. Some of the reasons are purely economic, some political, and some are a combination of the above. They include the rise in world population, the consolidation of agriculture into a massive, large-scale industry—here acquisitions and corporate consolidations are part of this pattern as well—and the role that chemicals have played in that consolidation process.

Beyond U.S. shores, internationally there have been

startling new alignments and interactions between European nations. These changes, global in their potential effects, will clearly reshape the lives of Europeans, Africans, Asians, and Americans into the year 2000 and beyond. The "opening up" of the Soviet Union, Eastern Europe, and East Germany, as their socialist system has declined, offers up unknown possibilities in the food trade as well as other commodities. Further, the consolidation of the Western European business communities from the older common market arrangement into new economic sectors of common currency exchanges in 1992 will begin to raise many questions about food production techniques and regulations. Oddly enough, the same factors which appear to foster the continuance of the chemical age in agriculture may also be the ones to turn it around. As the interest in clean food and water reverberates internationally, across national boundaries, as in the Green Party in Germany, the pressure to find safer alternatives in agriculture may produce a groundswell of public involvement and some positive results in the public and private agricultural sectors. An expanding global interest in saving the earth bodes well for citizens of the twenty-first century. The 1992 United Nations Global Conference on the Environment is evidence of it.

In a speech delivered by Mikhail Gorbachev, USSR President, to an international conference of scientists held in Moscow in 1990, he noted the following concerns for our environment:

> Great minds of the past foresaw the consequences of man's unthinking "subjugation" of nature. They warned: by destroying the plant and animal world and poisoning the soil, water, and air, the human race may destroy itself. By the end of the 20th century, we have the most acute crisis in relations between man, society, and nature.[3]

Speaking in language more closely resembling the writings of Rachel Carson than Karl Marx, Gorbachev points out that raising the standard of living should not be done at the expense of the environment:

> The greening of politics is a new view of the problem of consumption and its rationalization. Raising people's living standards must not take place at the cost of exhausting nature but should be accompanied by the conservation and renewal of the living conditions of the plant and animal world.
>
> The greening of politics also creates approaches to tackling many social tasks, connected primarily with damage to people's health as a result of the harm already done to the environment.[4]

If this speech is any indication of Mikhail Gorbachev's policies for the Soviet future, we can notice that the environment as a public concern is decidedly mainstream. Meanwhile, in Western Europe, environmental activists have already formed viable political parties. The Green Party founded in Germany will generate change from within the governmental system. When a society is involved in the active process of making change, that is the time many of its problems receive the most attention. We too can begin to address our environmental problems. One major concern that needs to be addressed is in the incredible rise in the numbers of people living on this planet—the population.

AGRICULTURE AND POPULATION— GETTING BIGGER ALL THE TIME

One change that challenges our present agricultural practices is the astounding growth in the world's popula-

tion. With population growth comes the concomitant demand for more food. The United Nations Development
Program tells us that the world population is now growing
by some 80 million people a year and is expected to reach
six billion by the end of the century. World population
reached five billion by mid-1987.[5] How can we produce
enough food to feed everyone, particularly when many
areas of the world are succumbing to drought and desertification. And some of that environmental damage is the
result of our twentieth-century agricultural practices.
Though not alone in this respect, some of our agricultural
technology has led to the depletion of soils and the pollution of watersheds, as well as the destruction of forests and
wildlife. So, as we survey the population explosion, it is
inevitable that the demand for food in the twenty-first
century will become one of the most compelling issues for
us to resolve.

AND THE WORLD IS GETTING SMALLER
ALL THE TIME

Agriculture on an international scale is becoming one
of the most important avenues of economic exchange. It all
began with the opening up of China for economic trade,
followed later by *glasnost* in the Soviet Union. In the concluding months of 1989, there was the breakup of the
Eastern European socialist block in Hungary, Czecho-
slovakia, and East Germany. Romania, with its singularly
bloody revolution, very nearly overshadowed the pro-
jected 1992 formation of an economically united European
community. The effect of these rapid and enormous changes
in Europe alone have yet to be understood, but certainly
international trade agreements are a critical factor in the

survival of these "new" economies. A key area in trade agreements is food. Exchanges in grain is nothing new in this century, but new interest is percolating over the possibilities of disseminating a host of fresh foods—produce which can be transported over extraordinarily long distances. For instance, in 1989, fresh corn grown in New Jersey fields was flown to Sweden to be consumed in restaurants only a day or two from the time of picking. Yet while all this excitement remains in the early stages of developing agreements, governments seem to be giving minimum thought to the problem of regulating chemicals in the growth of food. If a pesticide is outlawed in one country, but traded with another country where it is not, it will continue to affect the health of populations. So regulations within national boundaries, which are far from complete, appear to be already obsolete, undercut by the internationality of trading of food. While this is certainly not a new problem, it is a rapidly growing issue today that seems to have no mechanism for dealing with the question. We know, for example, that pesticides banned from use in the United States are shipped and sold overseas to be used in countries where no such regulations exist.

PESTICIDES: WHAT GOES AROUND COMES AROUND

Greenpeace, an independent environmental interest group that spans the continents, produced a U.S. report in 1989 titled "Exporting Banned Pesticides: Fueling the Circle of Poison." The report provided a case study of the environmental hazards and public health issues posed by a loophole in a federal law in FIFRA (Federal Insecticide,

Fungicide, and Rodenticide Act) in selling banned pesticides to other countries. In 1978, they inform us, EPA halted most agricultural uses of two hazardous and persistent pesticides, chlordane and heptachlor. "These organochlorine insectides are globally notorious for the harmful effects on public health and the environment left in the wake of their production, use and disposal."[6] To date, six U.S. states and at least 25 countries, including the European Economic Community, have banned or phased out chlordane, and 19 countries have banned most uses. Heptachlor has been banned totally in 12 countries and banned partly in 26. The report continues, however:

> Between 1987 and mid-1989, a chemical producing company shipped approximately 4,817,648 pounds of chlordane and heptachlor to 25 countries, with the majority of products going to or through, the Netherlands, Argentina, Australia and Singapore. Many of these countries are major trading partners of the U.S. . . . . However, the ultimate destination of the products is unknown to EPA, as well as to the Department of Agriculture and the FDA, who are responsible for monitoring imported foods for residues of pesticides.[7]

From this case study alone, we can begin to see the apparent crazy quilt pattern of regulating the health and safety of food. Banning the use of a particular pesticide is often a slow and tortuous task, and it occurs on a case-by-case basis when public awareness and concern create the climate for political action. Nevertheless, banning or even restricting the use of a chemical is hardly the end of its presence in our food and water. We do know that DDT is sold to Latin American countries where food is grown and exported to the United States. On top of this well-established trade relationship, evidence is mounting that new alliances in food trade will challenge the systems of regulation as they presently exist. It seems that correcting the problems left over from the twentieth century will keep us busy for some time to come. Rather than fantasizing about entirely new technologies, the people of the next millenium will no doubt find it necessary to turn back the clock to gain a perspective on their future life and to redirect their energies and practices toward contributing to an ecologically balanced world at large.

KEYS TO THE FUTURE

Improving our agricultural habits has to be the most important near-term future goal for the United States, as well as other agricultural producing countries. We do have other options, but we have simply failed to show an interest in their development until recently. Our government needs to find support for the research and utilization of integrated pest management (IPM) controls to replace our exclusive and heavy use of pesticides. IPM techniques will allow us to restrict chemical applications only to where absolutely needed and to implement biological controls, such as importing natural enemies of undesirable pests. We need to change food production practices which have damaged our environment for more than half of the twentieth century. These changes will not be easy to effect. Nevertheless, there are many keys that can be used to open the door to a healthier world:

1. The public has untapped power in the marketplace. The consumer can look for sources of organically grown or IPM-grown food products

and request that certification of these methods be regulated by state governments. The label "organically grown" must mean something.

2. The public can communicate with legislators on amending the loopholes in food regulation. The FIFRA needs shoring up, particularly with respect to the exporting of pesticides that cannot legally be used in the United States.

3. The Department of Agriculture needs to be supported financially and encouraged politically to pursue research in biological controls as effective alternatives to pesticide use. The Department also maintains a fine teaching unit, APHIS, the Animal and Plant Health Inspection Service. We need to assist farmers in learning how to use and adapt methodologies in the field. This too needs expansion of resources.

4. Farmers must be encouraged to refrain from pesticide use wherever and whenever possible. The change will be feasible only when the economics start to work in favor of healthier farming interests. An avenue of change from pesticide use should be introduced in areas where pesticides have already proven to be uneconomical. For example, Dr. William Bruckart of the U.S. Department of Agriculture's Agricultural Research Service (ARS) points out that weeds in the California rangeland cover far too extensive a tract of land to make the spraying of pesticides practical or economical. Yet the weed problem is enormous. Like insects, most weeds come from elsewhere and arrive without their natural enemies. Biological controls for the last 100 years have made use of

insects and with great success; it is only just recently that the ARS is trying to use fungi to control weeds. Pathogens, which are disease-causing viruses, are sought out in the same place as the offending weeds, and by manipulating the pathogen, it is possible to produce a bioherbicide. The leafy spurge in the northwest U.S. has been a terrible problem because of its poisonous effects on grazing cattle. A big priority for the ARS is to develop a biocontrol for control of those weeds. It is these kinds of innovations that utilize scientific knowledge to find solutions inherent in nature and which will eventually allow us to turn the tide on The Chemical Age.

That being said, the substantial use of pesticides becomes far less economically attractive to farmers.

Water protection though has a different set of problems. Food may be sensitive to the marketplace and offers the public a certain amount of leverage in changing agriculture, but water cannot be boycotted and is not a product of the consumer market. As such, there are only two ways to protect the quality of our water: (1) improve the regulating by state and federal governments in protecting the quality of our groundwater as well as surface water and (2) reduce the amount of chemical and natural pollutants used in our agricultural practices.

The goal of an ecologically healthy society necessitates some significant changes from the way we have lived our lives in the twentieth century. We cannot know yet what the face of the twenty-first century will look like, but we do know what we would like it to resemble. Certainly we need to call upon new technologies to help devise solutions for the many problems we have inadvertently created in the past and in the present. Let us not forget that biological controls are also a product of technological advances—that the use of nature to solve nature's problems is perhaps the highest form of technological advancement we can expect to achieve. In the final analysis, it is not technology that we abhor—but rather the misuse of it.

Notes

CHAPTER ONE

1. David Pimentel and John H. Perkins, eds., *Pest Control: Cultural and Environmental Aspects*, AAAS Selected Symposium (Boulder: Westview Press, 1980).
2. "Intolerable Risk: Pesticides in Our Children's Food," A report of the Natural Resources Defense Council, February 27, 1989.
3. John J. Schlebecker, *Whereby We Thrive: A History of American Farming, 1607–1972* (Ames: Iowa State University Press, 1975).
4. O. E. Rölvaag, *Giants in the Earth* (New York: Harper Brothers, 1927).
5. Rachel Carson, *Silent Spring* (Cambridge: Houghton Mifflin Co., 1962).
6. John J. Schlebecker, *Whereby We Thrive: A History of American Farming, 1607–1972* (Ames: Iowa State University Press, 1975).
7. John Steinbeck, *The Grapes of Wrath* (New York: Viking Press, 1939).
8. Rachel Carson, *Silent Spring* (Cambridge: Houghton Mifflin Co., 1962).

CHAPTER TWO

1. Carol B. Gartner, *Rachel Carson* (New York: Frederick Unger Publishing Co., 1983).

2. Ibid.
3. Rachel Carson, *Silent Spring* (Cambridge: Houghton Mifflin Co., 1962)
4. Ibid.
5. Ibid.
6. Ibid.
7. Ibid.
8. Carol B. Gartner, *Rachel Carson* (New York: Frederick Unger Publishing Co., 1983).
9. Paul Brooks, *The House of Life: Rachel Carson at Work* (Boston: Houghton Mifflin, 1972).
10. Ibid.
11. Rachel Carson, *Silent Spring* (Cambridge: Houghton Mifflin Co., 1962).
12. Ibid.
13. Ibid.
14. Betty Friedan, *The Feminine Mystique* (New York: W. W. Norton, 1963).
15. Paul Brooks, *The House of Life: Rachel Carson at Work* (Boston: Houghton Mifflin, 1972).
16. John Sheail, *Pesticides and Nature Conservation: The British Experience, 1950–1975* (Oxford: Clarendon Press, 1985).
17. Manfred Kroger, "The Why and How of Communicating Science," *Food Technology*, 41 (Jan. 1987), p. 93–99.
18. Joan Goldstein, *Environmental Decision Making in Rural Locales: The Pine Barrens* (New York: Praeger, 1981).
19. Bert L. Bohmont, *The New Pesticide Users' Guide* (Reston: Reston Publishing Co., 1987).
20. John J. Schlebecker, *Whereby We Thrive: A History of American Farming, 1607–1972* (Ames: Iowa State University Press, 1975).
21. Rachel Carson, *Silent Spring* (Cambridge: Houghton Mifflin Co., 1962).
22. John J. Schlebecker, *Whereby We Thrive: A History of American Farming, 1607–1972* (Ames: Iowa State University Press, 1975).
23. S. Gail Battista, "The Conviction of DDT," *Environmental Reporter*, 3/39, Jan. 26, 1973.
24. Rachel Carson, *Silent Spring* (Cambridge: Houghton Mifflin Co., 1962).
25. S. Gail Battista, "The Conviction of DDT," *Environmental Reporter*, 3/39, Jan. 26, 1973.

26. *New York Times*, "DDT: In the End The Risks Were Not Acceptable," June 18, 1972.

CHAPTER THREE

1. *Environmental Reporter*, Monograph #14, Vol. 3/39, January 26, 1973
2. Ibid.
3. *News Report* (Washington: National Academy of Sciences, XXI, No. 6, June–July, 1971).
4. Molly Joel Coye, "The Effects of Agricultural Production: I. The Health Effects of Agricultural Workers," *Journal of Public Health Policy*, Sept. 1985, pp. 349–370.
5. Molly Joel Coye, "The Health Effects of Agricultural Production: II. The Health of the Community," *Journal of Public Health Policy*, Autumn, 1986, pp. 340–354.
6. *Environmental Reporter*, Monograph #14, Vol. 3/39 January 26, 1973.
7. Thomas H. Moss and David L. Sills, eds., *The Three Mile Island Nuclear Accident: Lessons and Implications* (New York: Annals of the New York Academy of Science, Vol. 365, 1981).
8. "Intolerable Risk," A report of the Natural Resources Defense Council, 1989.
9. Congressional Research Services (CRS), "Pesticide Regulation: Legislative Debate about FIFRA in 1986," Washington, D.C., Library of Congress, May 11, 1987.
10. Ibid.
11. U.S. Congress House Committee on Government Operations. *Problems Plague the EPA's Pesticide Regulatory Activities*, Washington, 1984.
12. National Research Council, *Regulating Pesticides in Food: The Delaney Paradox*. (National Academy Press: Washington, 1987).
13. "Intolerable Risk," a report of the Natural Resources Defense Council, 1989.
14. *New York Times*, 5/16/89.
15. "Government Regulation of Pesticides in Food: The Need for Administrative and Regulatory Reform." Report by the Subcommittee on Toxic Substances, Environmental Oversight, Research

and Development to the Committee on Environment and Public Works, U.S. Senate, 1989.

16. Ibid.
17. Ibid.
18. Congressional Research Service (CRS), "Apple Alarm: Public Concern About Pesticide Residues in Fruits and Vegetables," Library of Congress, Environment and Natural Resources Policy Division, March 10, 1989.
19. Ibid.
20. Molly Joel Coye, "The Health Effects of Agricultural Production: II. The Health of the Community," *Journal of Public Health Policy*, Autumn, 1986, pp. 340–354.
21. "Intolerable Risks," A report of the Natural Resources Defense Council, 1989.
22. Congressional Research Service (CRS), "Apple Alarm: Public Concern About Pesticide Residues in Fruits and Vegetables," Library of Congress, March 10, 1989.

CHAPTER FOUR

1. Joan Goldstein, *Environmental Decision Making in Rural Locales: The Pine Barrens* (New York: Praeger, 1981).
2. Ted Conover, *Coyotes* (New York: Vintage Books, 1987).
3. Molly Joel Coye, John A. Lowe, and Keith J. Maddy, "Biological Monitoring of Agricultural Workers Exposed to Pesticides: II. Monitoring of Intact Pesticides and Their Metabolites," *Journal of Occupational Medicine*, Vol. 28, No. 8, Aug., 1986, pp. 629–636
4. John Steinbeck, *The Grapes of Wrath* (New York: Viking Press, 1939).
5. Michael O'Malley, "Recognizing Fieldworker Poisoning," *Migrant Health Clinical Supplement*, June/July, 1988.
6. Ibid.
7. Keith T. Maddy and Susan Edminston, "Summary of Illnesses and Injuries Reported in California by Physicians in 1986 as Potentially Related to Pesticides," Report of the California Department of Food and Agriculture, Division of Pest Management,

Environmental Protection and Worker Safety, Sacramento, Ca., October 5, 1987.

8. Michael O'Malley, "Priority Investigations Involving Phosalone in Fresno and Madera Counties," 1987, California Department of Food and Agriculture.

9. Molly Joel Coye et al., "Clinical Confirmation of Organophosphate Poisoning of Agricultural Workers," *American Journal of Industrial Medicine*, 10:399–409, 1986.

10. Michael O'Malley, "Recognizing Fieldworker Poisoning" *Migrant Health: Clinical Supplement*, June/July, 1988.

11. L. Duncan Saunders, Richard G. Ames *et al.*, "Outbreak of Omite-CR Induced Dermatitis Among Orange Pickers in Tulare County, Ca.," *Journal of Occupational Medicine*, Vol. 29, No. 5, May 1987, pp. 409–413.

12. Michael O'Malley, "Recognizing Fieldworker Poisoning," *Migrant Health Clinical Supplement*, June/July, 1988.

13. Ibid.

14. Stephanie K. Brown, Richard G. Ames *et al.*, "Occupational Illness from Cholinesterase Inhibiting Pesticides Among Agricultural Workers in California," *Archives of Environmental Health*, Vol. 44, No. 1, Jan/Feb. 1989, pp. 34–39.

15. Margaret E. Scarborough, Richard G. Ames *et al.*, "Acute Health Effects of Community Exposure to Cotton Defoliants," Hazard Evaluation Section Office of Environmental Health Hazard Assessment, California Department of Health Services, Berkeley, March, 1989.

16. Molly Joel Coye, Paul G. Barnett *et al.*, "Clinical Confirmation of Organophosphate Poisoning by Serial Cholinesterase Analysis," *Archives of Internal Medicine*, Vol. 147, March 1987, pp. 438–442.

17. Keith T. Maddy and Susan Edmiston, "Summary of Illnesses and Injuries Reported in California by Physicians in 1986 as Potentially Related to Pesticides," Report of the California Department of Food and Agriculture, October 5, 1987.

18. Ibid.

19. Bert L. Bohmont, *The New Pesticide Users Guide* (Reston: Reston Publishing Co., 1987).

20. Molly Joel Coye, "The Health Effects of Agricultural Production: The Effects on Agricultural Workers," *Journal of Public Health Policy*, Vol 6, No. 3, Sept. 1985.

21. Bert L. Bohmont, *The New Pesticide Users Guide* (Reston: Reston Publishing Co., 1987).

22. Cheryl Best, "Natural Pest Controls," *Garbage*, Sept./Oct. 1989.

23. L. Duncan Saunders, Richard G. Ames *et al.*, "Outbreak of Omite-CR Induced Dermatitis Among Orange Pickers in Tulare County California," *Journal of Occupational Medicine*, Vol 29, No. 5, May, 1987, pp. 409–413.

CHAPTER FIVE

1.   Joan Goldstein, *Environmental Decision Making in Rural Locales: The Pine Barrens* (New York: Praeger, 1981).

2.   John McPhee, *The Pine Barrens* (New York: Ballantine Books, 1967).

3.   *Alternative Agriculture* (National Research Council, National Academy Press, Washington, D.C., 1989).

4.   Ibid.

5.   Ibid.

6.   Congressional Research Service (CRS), "Groundwater Quality Protection: Issues in the 101 Congress, Library of Congress, June, 1989.

7.   Congressional Research Service (CRS), "Groundwater Quality: Current Federal Programs and Recent Congressional Activities, Library of Congress, March 1, 1989.

8.   Congressional Research Service (CRS), "Groundwater Quality Protection: Issues in the 101 Congress, Library of Congress, June, 1989.

9.   Joan Goldstein, "Planning for Women in the New Towns: New Concepts and Dated Roles" *The Journal of Comparative Family Studies*, Vol. IX, No. 3 (Autumn, 1978), pp. 385–392.

10. "Sampling for Pesticide Residues in California Well Water, 1988 Update." Annual Report to the Legislature, State Department of Health Services, Dec. 1, 1988.

11. Congressional Research Service (CRS), "Groundwater Quality Protection: Issues in the 101 Congress", Library of Congress, June, 1989.

12. *Alternative Agriculture* (National Research Council, National Academy Press, Washington D.C., 1989).

13. Congressional Research Service (CRS), "Groundwater Quality: Current Federal Programs and Recent Congressional Activities," Library of Congress, March 1, 1989.
14. Congressional Research Service (CRS), "Groundwater Contamination and Protection, Update, December 27, 1988," Library of Congress.
15. *Clean Water Action News*, Fall, 1988.

CHAPTER SIX

1. David Pimentel and John H. Perkins, eds., *Pest Control: Cultural and Environmental Aspects*, AAAS Select Symposium (Boulder: Westview Press, 1980).
2. Ibid.
3. Ibid.
4. Ibid.
5. Agricultural Research (USDA), "Alternative Versus Conventional Farming," ARS, Oct., 1989.
6. John and Allan A. Lomax, *American Ballads and Folk Songs* (New York: The MacMillan Co., 1934).
7. Agricultural Research, ARS, March, 1989.
8. Ibid.
9. *New York Times*, "Medflies and Malathion," February 28, 1990, p. A26.
10. Ibid.
11. Robert P. Kahn, ed., *Plant Protection and Quarantine*, Vol. 111, Special Topics (Boca Raton: CRC Press Inc., 1987).
12. Jack R. Coulson and Richard S. Soper, "Protocols for the Introduction of Biological Control Agents in the U.S.," in *Plant Protection and Quarantine*, Vol. III, Special Topics (Boca Raton: CRC Press Inc., 1987).
13. Ibid.
14. Agricultural Research (USDA), "Alternative Versus Conventional Farming," ARS, Oct. 1989.
15. Ibid.
16. International Symposium on Biological Control Implementation, McAllen, Texas, April 4–6, 1989.

CHAPTER SEVEN

1. Speech by Gaylord Nelson, Earth Day Committee. *Earthline,* Earth Day 1990, Issue #1.
2. Speech by Denis Hayes, Earth Day Committee. *Earthline,* Earth Day 1990, Issue #1.
3. Jon Kerner and Kurt Fensterbush, "Profiles of Toxic Waste Victims' Movements for Major Sites on the National Priorities List." Paper presented at Southern Sociology meetings, 1988.
4. Ibid.
5. Mothers and Others, *TLC Newsletter,* Vol. 1, No. 1, Fall, 1989.
6. *New York Times,* "Apple Chemical Being Removed in U.S. Market," June 3, 1989.
7. *Supermarket News,* Vol. 39, No. 28, July 10, 1989.
8. Ibid.
9. Ibid.
10. Ibid.

CHAPTER EIGHT

1. The American Council on Science and Health is a largely industry-backed organization with support from the chemical industry.
2. The Center for Communication, New York City, is an educational organization devoted to studying the role of media.
3. Hodding Carter III, "Alar Scare: Case Study in Media Skewed Reality," *Wall Street Journal,* April 20, 1989.
4. "Alar Fears Unfounded," *New York Times,* October 15, 1989.
5. Ibid.
6. "Tight Limits Proposed for Popular Farm Chemical," *New York Times,* December 4, 1989.
7. Reporting on the Environment: Are We Scaring Ourselves to Death? A forum presented at The Center for Communication, New York City, October 26, 1989.
8. Profile: "Meet ACSH Vice President, Edward G. Remmers, Sc.D.," *ACSH News and Views,* May–June, 1987.
9. Manfred Kroger, "The Why and How of Communicating Science," *Food Technology,* 41(1), 93–99, 1987.

10. Ibid.
11. Ibid.
12. Ibid.
13. Ibid.
14. Ibid.
15. Elizabeth M. Whelan, "A Morbid Fear of Illness Makes America Trash Good Food and Common Sense," *Los Angeles Times*, March 20, 1989.
16. Mary Painter, *Nutrition Watch*, 1988.
17. Michael Greenberg *et al.*, "Network Evening News Coverage of Environmental Risk," *Risk Analysis*, Vol. 9, No. 1, 1989.
18. Henry Fairlie, "Fear of Living: America's Morbid Aversion to Risk," *New Republic*, January 23, 1989.
19. Ibid.
20. Peter Passell, "Life's Risks: Balancing Fear Against Reality of Statistics," *New York Times*, May 8, 1989.
21. Ibid.

CHAPTER NINE

1. Henry Fairlie, "Fear of Living: America's Morbid Aversion to Risk," *New Republic*, January 23, 1989.
2. Ibid.
3. Mikhail Gorbachev, speech presented at the Global Forum on the Environment, Moscow, January 20, 1990, as reported in Pravda, p. 6.
4. Ibid.
5. United Nations report, "A Better Environment for Development," United Nations Development Program.
6. "Exporting Banned Pesticides: Fueling the Circle of Poison," A Greenpeace Report, August, 1989.
7. Ibid.

Index